LESSONS FROM DISASTER

How Organizations Have No Memory
and Accidents Recur

Trevor Kletz

INSTITUTION OF CHEMICAL ENGINEERS

The information in this book is given in good faith and belief in its accuracy, but does not imply the acceptance of any legal liability or responsibility whatsoever, by the Institution, or by the author, for the consequences of its use or misuse in any particular circumstances.

All rights reserved. No part of this publication may be reproduced, stored in a retrieval system, or transmitted, in any form or by any means, electronic, mechanical, photocopying, recording or otherwise, without the prior permission of the copyright owner.

Published by
Institution of Chemical Engineers
Davis Building
165–171 Railway Terrace
Rugby, Warwickshire CV21 3HQ, UK.

Copyright © 1993 Institution of Chemical Engineers

ISBN 0 85295 307 0

Printed in the United Kingdom by Redwood Press Limited, Melksham, Wiltshire.

PREFACE

I owe a great debt to the many colleagues, in industry and elsewhere, who have supplied the information and ideas discussed in this book. I could not have written it without their help. Many of these contributions were made during the discussions described in Section 10.2 and the title of Chapter 2 is based on a comment made (I do not know by whom) at one of them. I also owe a debt to the companies who have allowed me to describe the accidents they have had, and a special word of thanks is due to the senior managers of ICI, at the time I worked there, for their willingness to let me describe some of the company's accidents. They set an example to the industry that is not yet followed as widely as it should be (see Chapter 5). However, the accidents described in the following pages occurred in many different companies. I would also like to thank the Department of Chemical Engineering at Loughborough University of Technology, and especially Professor Frank Lees, for giving me the opportunity to carry on with my work after I retired from industry.

To avoid the clumsy phrases 'he or she' and 'him or her' I have used 'he' or 'him'. There has been a welcome increase in the number of women employed in industry but the manager, designer or accident victim is still usually male.

Some of the chapters are based on papers that I have already published and thanks are due to the American Institute of Chemical Engineers for permission to quote from the following:
Loss Prevention, 1980, 13: 1 (in Chapter 2)*;
Loss Prevention, 1976, 10: 151 (in Chapter 3)*;
Plant/Operations Progress, July 1988, 7 (3): 145 (in Chapter 5)*.

Finally, I would like to thank Jim Watson for preparing the figures and the editorial staff of the Institution of Chemical Engineers for their help and patience.

---

*   Used by permission of the American Institute of Chemical Engineering © AIChE.

A NOTE FOR AMERICAN READERS

The term 'plant manager' is used in the UK sense to describe the first level of professional management, someone who would be known as a supervisor in most US companies. The person in charge of a site is called a works manager.

REFERENCES

The Library and Information Service of the Institution of Chemical Engineers in Rugby, UK, offers a worldwide service for the supply of the references listed in this book.

CONTENTS

PREFACE		iii
FORETHOUGHTS		viii
1.	**INTRODUCTION**	1
2.	**ORGANISATIONS HAVE NO MEMORY**	4
2.1	Isolation for Maintenance	4
2.2	Explosions in Buildings	10
2.3	Choked Vents	15
2.4	Pressure Vessels Opened by Operators	18
2.5	Improving the Corporate Memory	21
APPENDIX 2.1 — 'IT MUST NEVER HAPPEN AGAIN'		24
3.	**ACCIDENTS OF THE COMING YEAR**	27
3.1	A Tank will be Sucked in	27
3.2	A Trip will Fail to Operate	29
3.3	A Road Tanker will be Overfilled	31
3.4	A Man will be Injured while Disconnecting a Hose	31
3.5	A Road or Rail Tanker will be Moved before the Hose has been Disconnected	32
3.6	The Wrong Pipeline will be Opened	33
3.7	A Heavy Oil Tank will Foam Over	34
3.8	A Pipe will be Damaged by Water Hammer	35
3.9	Must these Incidents Occur Again?	36
APPENDIX 3.1 — HOW STRONG IS A STORAGE TANK?		37
4.	**SOME OTHER FAILURES TO LEARN FROM THE PAST**	41
4.1	Knowledge in the Wrong Place	41
4.2	Other Examples of Forgotten Knowledge	45
4.3	Loss of Skill	49
4.4	Two Fires and an Explosion in an Oil Company	53
4.5	Some Incidents on the Railways	56
4.6	Some Successes	58

5.	**WHY ARE WE PUBLISHING FEWER ACCIDENT REPORTS?**	62
5.1	THE PRESENT POSITION	62
5.2	WHY SHOULD WE PUBLISH ACCIDENT REPORTS?	63
5.3	WHY ARE WE PUBLISHING FEWER ACCIDENT REPORTS?	64
5.4	HOW CAN WE ENCOURAGE THE TRANSFER OF INFORMATION?	67
6.	**WHAT ARE THE CAUSES OF CHANGE AND INNOVATION IN SAFETY?**	70
6.1	ISOLATION FOR MAINTENANCE	70
6.2	TRIP AND ALARM TESTING	74
6.3	STORAGE OF LIQUEFIED PETROLEUM GAS	76
6.4	HIGH INTEGRITY PROTECTIVE SYSTEMS	81
6.5	INHERENTLY SAFER DESIGN — A SLOWER CHANGE	83
6.6	OTHER SPRINGS OF ACTION	85
6.7	HOW CAN WE ENCOURAGE CHANGE AND INNOVATION?	88

APPENDIX 6.1 — A NOTE ON ASYMPTOTES 92

7.	**THE MANAGEMENT OF SAFETY**	93
7.1	POLICY	93
7.2	WHOSE RESPONSIBILITY?	94
7.3	TRAINING, INSTRUCTIONS AND LEARNING THE LESSONS OF THE PAST	96
7.4	AUDITS AND THEIR LIMITATIONS	96
7.5	TESTING AND INSPECTION	101
7.6	IDENTIFYING AND ASSESSING HAZARDS	102
7.7	ACCIDENT INVESTIGATION AND HUMAN ERROR	106
7.8	OTHER KEY PROCEDURES	109
7.9	THE OBVIOUS	114
7.10	THE SPECIAL RESPONSIBILITIES OF THOSE AT THE TOP	115

APPENDIX 7.1 — A FUTURE ACCIDENT REPORT 120

APPENDIX 7.2 — TRAIN DRIVER ERROR 122

APPENDIX 7.3 — A NOTE ON SPARK-RESISTANT TOOLS 124

8.	**AN ANTHOLOGY OF ACCIDENTS**	125
8.1	RELUCTANCE TO RECOGNISE THAT SOMETHING IS WRONG	125
8.2	LABELLING SHOULD BE CLEAR AND UNAMBIGUOUS	127
8.3	REASSESS HAZARDS FROM TIME TO TIME	127
8.4	WATCH CONTRACTORS CLOSELY	128

8.5	Do not Overload Supervisors	131
8.6	Do not Change a Design without Consulting the Designer	132
8.7	What should we do with Safety Equipment that Stops us Running the Plant?	133
8.8	Not all Trips are Spurious	134
8.9	'It must be Safe as I'm Following all the Codes'	136
8.10	Little-Used Equipment becomes Forgotten	136
8.11	Crowd Control	137
8.12	Different People — Different Views	138
8.13	We can Learn from a Report even if its Conclusion is Wrong	140
9.	**CHANGES IN SAFETY — A PERSONAL VIEW**	143
9.1	Early Days	143
9.2	Changing Responsibilities	145
9.3	Safety Adviser	149
10.	**IMPROVING THE CORPORATE MEMORY**	166
10.1	Spreading the Message	166
10.2	Discussions are Better than Lectures	167
10.3	Remembering the Message	170
10.4	Finding Old Reports	171
10.5	A Final Note	173

AFTERTHOUGHTS 175

INDEX 176

FORETHOUGHTS

'Forgetfulness is one of the great sins of our time. People block out remembrance of difficult times, of failures, of their own weakness.'
Albert Friedlander

'Wise men profit more from fools than fools from wise men; for the wise shun the mistakes of fools, but fools do not imitate the successes of the wise.'
Marcus Cato (234–149 BC), quoted by Plutarch, *Lives*

'Good counsel failing men may give, for why,
He that's aground knows where the shoal doth lie.'
Benjamin Franklin (1706–1790), *Poor Richard's Almanac*, 1758

'Experience is the best of schoolmasters, only the school-fees are heavy.'
Thomas Carlyle (1795–1881), *Miscellaneous Essays*, 1: 137

'It should not be necessary for each generation to rediscover principles of safety which the generation before already discovered. We must learn from the experience of others rather than learn the hard way. We must pass on to the next generation a record of what we have learned.'
Jesse C. Ducommun

'In biology both goods and messages are passed on from one generation to the next. But it is the messages that are the most important inheritance: only they can persist over millions and millions of years.'
A.G. Cairns-Smith, 1985, *Seven Clues to the Origin of Life* (Cambridge University Press), 12

1. INTRODUCTION

'What has happened before will happen again. What has been done before will be done again. There is nothing new in the whole world.'
Ecclesiastes, 1:9 (*Good News Bible*)

It might seem to an outsider that industrial accidents occur because we do not know how to prevent them. In fact, they occur because we do not use the knowledge that is available. Organisations do not learn from the past or, rather, individuals learn but they leave the organisation, taking their knowledge with them, and the organisation as a whole forgets.

This book describes many accidents that have been repeated, often many times, mainly, but not entirely, in the process industries. It also suggests some ways of improving the corporate memory. Though intended primarily for managers, designers and safety advisers, it will also interest students of the ways organisations work, or rather fail.

Though the main theme of the book is the importance of learning and remembering the lessons of the past (we should follow the example of Lot's wife and Orpheus and look back, but in our case to avoid disaster) I also discuss many of these lessons. The method I have adopted is (to quote S.J. Gould) a mixture of particulars that will, I hope, fascinate and generalities that will, I hope, instruct[1].

We cannot make gold from lead because there is no known process. We can, however, make our plants safer, if we wish to do so, by using information present in the open literature (including this book) and our company files and information gleaned by talking to our colleagues.

We spend much time and money acquiring new knowledge and describing the results at conferences and in journals. We spend much less effort answering the question, 'How can we persuade people to use the knowledge that is already available?' The knowledge may not be known to us but it is known to other people.

During my fourteen years as a safety adviser with ICI I was often tempted to tell a manager who had allowed a familiar accident to happen on his plant, 'Don't bother to write the accident report. I'll send you a copy from my files'.

On several occasions I have made myself unpopular, at a meeting called to discuss a proposed programme of research on some aspect of safety, by asking, 'How are we going to persuade people to use the new knowledge we are going to give them, when they are not using the knowledge we already have?'.

Nevertheless, I do not wish to disparage or discourage those who spend their days researching on safety. There is still much we need to know, and these researchers fulfil an important and necessary task. But I do suggest that we spend on methods of improving the corporate memory at least a small fraction of the time and money we spend on discovering new things to remember.

This failure to use the knowledge on our doorsteps is not, of course, restricted to industry or engineers. In a fascinating book, *Wonderful Life*[2], S.J. Gould shows how palaeontology had been revolutionised by the study of fossils that had been lying in museum drawers for decades; similarly, new light is being thrown on the society in which Christianity developed by the study of fragments of the Dead Sea Scrolls that scholars have been sitting on for forty years. However, in those cases fossils had to be examined or manuscripts deciphered. Information on accidents that have happened and the action needed to prevent them happening again is available to all who can read.

In *History — Remembered, Recovered, Invented*[3], Bernard Lewis distinguishes three sorts of history. In 'History Remembered', knowledge of the past has been passed on without a break by the written word or by word of mouth. The more recent accidents described in Chapter 2 and later chapters are 'History Remembered'. Most of the people on the plants concerned have not (yet?) forgotten them and if they have not read the reports they have heard about them from others.

In 'History Recovered', knowledge of events that has been completely forgotten is brought back to light by archaeologists or by the discovery of ancient documents. Some of the incidents described in Chapter 2 (Section 2.1, Incident 1 and Section 2.3, Incident 1, for example) are 'History Recovered'. The first example had been completely forgotten until I found the report while searching old files; the second had been almost forgotten but someone who had been present at the time vaguely remembered it and looked for the report. In industry ancient documents are those more than ten years old.

In 'History Invented', history is rewritten or invented to suit political ends or destroy unpleasant memories. The past is described, not as it was, but as the writer thinks it ought to have been. Though the rewriting of history is a flourishing industry in many parts of the world (including the West), there are, I hope, no examples of 'History Invented' in this book. We should, however, always be on our guard against the temptation to select from the past only those facts that support our views. History can be invented by omission and selection as well by deliberate deceit. The facts of history are those that historians have selected for scrutiny[4].

Early societies saw history as cyclical; like the seasons, everything comes round again in time. The Bible was the first book, the Israelites the first people, to see history as linear, as movement towards a goal, and in the West we have come to accept this as normal, or at least as desirable. The cyclical view, however, must correspond to something deep in human experience or we would not have believed in it for so long. Accidents do come round again every few years but no law of nature says they must do so. If we want to, we can break out of the cycle.

Chapter 2 describes four serious accidents that occurred in the United Kingdom chemical industry and were repeated ten or more years later in the same company. Chapter 3 describes some less serious accidents that are forgotten sooner and repeated more frequently while Chapter 4 describes some other failures to learn from the past. Chapter 5 tackles the reasons why we should, but do not, publish more accident reports. In Chapter 6 there is a change of theme to the causes of change and innovation in safety, while the management of safety is dealt with in Chapter 7. Chapter 8 returns to the main theme and describes some common but relatively simple accidents in the hope that they will stick in the memory. Chapter 9 shows how I came to be interested in safety and describes some incidents in which I was involved. Finally, Chapter 10 brings together and discusses further the main recommendations made in earlier chapters.

As in my other books, there are a number of quotations. I have not included them to claim the support of the famous but because they say something better than I can. The essence of a good quotation is that the reader says, 'That's what I've always thought, but I've never been able to put it so clearly'. Whenever possible I have checked my quotations; many popular versions of them are incorrect, or attributed to the wrong person[5].

You may not agree with all my recommendations. If so, please decide what you intend to do instead. Please do not ignore the incidents. They will happen yet again unless we take action to prevent them happening.

REFERENCES IN CHAPTER 1
1. Gould, S.J., 1986, *The Flamingo's Smile* (Penguin Books, London, UK), Prologue.
2. Gould, S.J., 1989, *Wonderful Life* (Norton, New York, USA).
3. Lewis, B., 1975, *History — Remembered, Recovered, Invented* (Princeton University Press, Princeton, New Jersey, USA).
4. Carr, E.H., 1990, *What is History?*, 2nd edition (Penguin Books, London, UK).
5. Boller, P.F. and George, J., 1989, *They Never Said It* (Oxford University Press, New York, USA).

2. ORGANISATIONS HAVE NO MEMORY

'Most discoveries are made regularly every fifteen years.'
George Bernard Shaw, 1906, *The Doctor's Dilemma*

This chapter describes four serious accidents in the UK chemical industry which were repeated ten or more years later in the same company, though not always in the same part of it. With the passage of time and changes in staff, the recommendations made after the original accidents were forgotten or had, perhaps, not been passed on to other parts of the organisation. Organisations have no memory; only people have memories and they move on.

After about ten years most of the staff on a plant have changed. No-one can remember why equipment was installed or procedures introduced and someone keen to cut costs or improve efficiency, both very desirable things to do, asks 'Why are we following this time-consuming procedure?' or 'Why are we using this unnecessary equipment?' No-one can remember and the procedures may be abandoned or the equipment removed.

The incidents, or others like them, were reported in the open literature (or in company reports widely circulated to other companies) but nevertheless they have continued to happen again elsewhere. I shall suggest some ways of improving the corporate memory.

2.1   ISOLATION FOR MAINTENANCE

INCIDENT 1
In 1928 a 36 inch (0.91 m) diameter low pressure gas line was being modified and a number of joints had been broken. Before work started the line was isolated by a closed isolation valve from a gasholder containing hydrogen, swept out with nitrogen and tested to confirm that no flammable gas was present. Unknown to the men on the job, the isolation valve was leaking. Eight hours after the job started the leaking gas ignited, there was a loud explosion and flames appeared at a number of the flanged joints on the line (Figure 2.1). One man was killed, by the pressure, not the flames, but damage was slight.

The source of ignition was a match, struck by one of the workmen near an open end so that he could see what he was doing. He thought it was safe to strike a match as he had been assured that all flammable gas had been removed

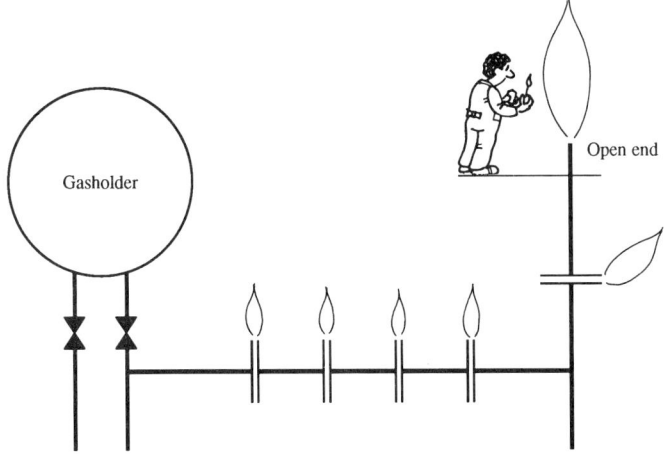

Figure 2.1 Although the pipelines had been swept out, when a man struck a match near the open end, an explosion occurred and flames appeared at the flanged joints. The gasholder valve was leaking.

from the plant. Once a flammable mixture is formed a source of ignition is always liable to turn up, so the real cause of the explosion was not the match but the leaking valve. The match was merely the triggering event.

The report on the accident made three recommendations:
(1) Never trust an open gas main which is attached to a system containing gas, and keep all naked lights clear.
(2) When working on pipebridges at night, adequate lighting should be available.
(3) Never place absolute reliance on a gasholder valve, or any other valve for that matter. A slip-plate is easy to insert and absolutely reliable.

The third recommendation was repeated in a Safety Handbook given to every employee the following year. (Figure 2.2, taken from this Handbook, shows equipment similar to that involved in the accident — see page 6.)

Over the years this sound advice was forgotten and the use of slip-plates (blinds) for isolating equipment under repair was allowed to lapse (except for vessel entry). We do not know why. Perhaps it was difficult to insert slip-plates because the plants were not designed to take them. Perhaps, with changes in

5

Figure 2.2 Gas leaked through a closed valve and was ignited at an open end similar to this one (bottom left).

staff, the accident was forgotten and slip-plating abandoned to save time and trouble. Meanwhile the company expanded, the site became full and a new site was developed 10 km away.

INCIDENT 2

In 1967, on the new site, a large pump was being dismantled for repair. When a fitter removed a cover, hot oil came out and caught fire as the 14 inch (0.36 m) diameter pump suction valve had been left open. The temperature of the oil was about 280°C, above its auto-ignition temperature, so ignition was quick and inevitable. Three men were killed and the unit, only a year old, was destroyed. One of the men killed was an electrician who was rigging up temporary lights so that work could continue after dark, so it seems that the second 1928 recommendation had not been forgotten.

We do not know why the suction valve had been left open. The foreman who signed the permit-to-work said that he inspected the pump before issuing the permit and found the valve already closed. Either his recollection is incorrect or, after he inspected the pump, someone opened the valve. There was no lock, tag or slip-plate to prevent him doing so.

After the fire, instructions were issued that before any equipment is handed over to the maintenance organisation:
(1) The equipment must be isolated by slip-plates (blinds) (or physical disconnection and blanking) unless the job to be done is so quick that fitting slip-plates (or disconnecting) would take as long as the main job and be as hazardous.
(2) Valves used to isolate equipment for maintenance, including isolation for slip-plating (or disconnection), must be locked shut with a padlock and chain or an equally effective device.
(3) If there is a change in intention — for example, if it is decided to dismantle a pump and not just work on the bearings — the permit-to-work must be handed back and a new one issued.

The second incident, discussed further in Section 6.1 on page 70, had much more serious consequences than the first one as the material involved was a hot liquid instead of a cold low-pressure gas.

INCIDENT 3
The second incident was so traumatic that it was not forgotten on the site where it occurred. But it was forgotten on the main site, the site where the first incident had occurred. In 1987 a hydrogen line, about 12 inches (0.3 m) diameter, had to be repaired by welding. The hydrogen supply was isolated by closing three valves in parallel (and a fourth valve in series with two of them; see Figure 2.3 on page 8). The line was then purged with nitrogen and tested at a drain point before welding started, to confirm that no hydrogen was present. When the welder struck his arc an explosion occurred, and he was injured. The investigation showed that two of the isolation valves were leaking. It also showed why hydrogen was not detected at the drain point: it was at a low level and air was drawn through it into the plant to replace gas leaving through a vent (see Figure 2.3 on page 8). The source of ignition was sparking because the welding return lead was not securely connected to the plant (another familiar problem)[1].

The accident report recommended that slip-plates should be used in future for the isolation of hazardous materials. There was no reference to the earlier incidents which were probably not known to the author.

7

LESSONS FROM DISASTER

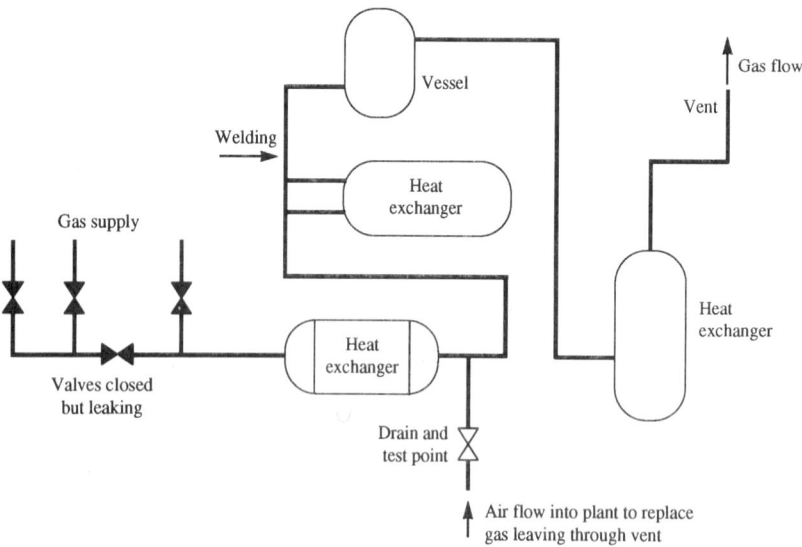

Figure 2.3   Simplified diagram of plant showing why hydrogen gas was not detected at the drain point.

Figure 2.4   Damage caused by explosion at the Phillips polyethylene plant in Pasadena, Texas, in 1989. (Photo: AP.)

INCIDENTS IN OTHER COMPANIES

In 1989 an explosion at the Phillips high-density polyethylene plant in Pasadena, Texas killed 23 employees and injured over 130 (see Figure 2.4). Damage amounted to $750m, debris was thrown six miles and the subsequent fire caused two iso-butane tanks to BLEVE, that is, the heat weakened the metal of the tank walls so that they burst (see Section 6.3, page 76). The initial explosion occurred less than two minutes after the start of an ethylene leak but nevertheless 40 tonnes were involved and the explosive force was equivalent to 2.4 tonnes TNT.

The immediate cause of the leak was simple: a length of pipe was opened up to clear a choke while the isolation valve (which was operated by compressed air) was open. To quote the official report[2], 'Established Phillips corporate safety procedures and standard industry practice require back-up protection in the form of a double valve or blind flange insert whenever a process or chemical line in hydrocarbon service is opened. Phillips, however, at the local level, had implemented a special procedure for this maintenance operation which did not incorporate the required backup'.

The air hoses which supplied power to the valve had been accidentally connected up the wrong way round; as a result the valve was open when its actuator was in the closed position. Identical couplings were used for the two connections so it was easy to reverse them. According to company procedures they should have been disconnected during maintenance, but they were not. The valve could be locked open or closed but the lock was missing.

The year before the Phillips explosion 167 people were killed in the Piper Alpha explosion and fire. According to the official report[3], a pump relief valve was removed for overhaul and the open end blanked. Another shift, not knowing the relief valve was missing, started up the pump (Paragraph 6.109). The blank was probably not tight and condensate (that is, light oil) leaked past it and exploded. Lord Cullen's report said, ' ... I have come to the conclusion that on a balance of probabilities the leakage of condensate was from a blind flange assembly which was not leak-tight' (Paragraph 6.187).

After describing other features of the permit-to-work system the report said, 'These examples serve to demonstrate that the operating staff had no commitment to working to the written procedure; and that the procedure was knowingly and flagrantly disregarded' (Paragraph 11.3). Paragraph 18.29 recommended some changes to the procedure.

The results of the errors on Piper Alpha were more serious than those of the other incidents I have described because of the congestion characteristic of an oil platform which made escape difficult . The official report has much to say on ways of mitigating these factors (Chapters 19 and 20).

Another leak on an offshore oil platform had a similar cause, although this time it did not ignite. According to one report[4], a flange was removed from a relief pipe connected to the flare system. 'The permit-to-work system should have ensured that the work was authorised and the work site made safe, preferably by locked valves, before the work was started.'

The nuclear industry has also learned by experience that they should not rely on valve isolations. One of the few criticality incidents that have occurred was the result of doing so. At Oak Ridge, Tennessee, in 1958 some small storage vessels were disconnected from the plant for cleaning and then reconnected. The next step was to fill them with water for leak testing and then drain the water into a 200 litre drum. Before this could be done, some enriched uranium solution entered the vessels through a leaking valve; it was drained into the drum ahead of the water. The vessels were shaped so that they could not go critical but the drum was not, and a criticality incident — a burst of neutrons — occurred[5].

2.2 EXPLOSIONS IN BUILDINGS

INCIDENT 1
In 1956 a leak of propylene occurred from the gland of an injector, a high pressure reciprocating pump, operating at a pressure of 250 bar. The leak was due to the failure by fatigue of the studs holding the gland in position. The injector was located in an unventilated building; escaping liquid vaporised, escaped through a large open door and was ignited by a furnace 75 m away, a few minutes after the leak started. Four men were badly burned. One of them was putting on breathing apparatus with the intention of entering the building to isolate the leaking injector and the others were assisting him. Damage to the building was slight, mainly broken windows.

The injector (and several others) had originally been used for pumping hydrocarbons similar to petrol but they were no longer required on this duty and were used to pump propylene. The vapour from a leak of petrol will not travel anything like 75 m but no-one realised that the vapour from a leak of liquefied gas can travel hundreds of metres. The underlying cause of the fire was thus the inexperience of the managers and designers, who were familiar with the properties of petrol but not with those of liquefied petroleum gases (LPG) such as propylene. The mechanical suitability of the injector for the new duty was checked but no-one looked into wider questions such as the behaviour of leaks.

After the fire a considerable sum of money was spent on improving the plant:

- The injectors were resited in an open-sided building (really just a canopy), so that small leaks would be dispersed by natural ventilation. (Indoors a few tens of kilograms are sufficient to destroy a building; in the open air several tonnes are necessary.)
- The injectors and associated equipment were surrounded by a steam curtain, to prevent leaks reaching a source of ignition.
- Flammable gas detectors were installed so that leaks would be detected promptly.
- Remotely operated valves were installed so that leaking injectors could be isolated and blown down, quickly, from a safe distance. (Eight years later, when another leak occurred, the foremen forgot that they were there. See Section 8.10, page 136.)
- A flare system was installed so that relief valve discharges and deliberate leaks, necessary when preparing equipment for maintenance, did not have to be discharged to atmosphere.
- The LPG storage vessels were relocated well away from operating plant.
- Changes were made to the gland retaining mechanism on the injectors. They are described in Section 4.3.1 on page 49.

By reusing old equipment the original plant had been built cheaply. It was not so cheap in the end.

For the next twelve years the company located all LPG equipment in the open air or in open-sided buildings. In 1968 the site chosen for a new compressor house was close to a workshop. In order to reduce the noise level in the workshop the building was completely enclosed. The explosion had been forgotten and the noise problem loomed large. The building had just been completed when the report on the explosion described below was issued. The men who had built the compressor house walls were asked to remove them. Their comments have not been preserved.

INCIDENT 2

The company which suffered this incident was under the same ownership as that involved in the last one, and was not far away, but its technical management was quite independent.

A leak of ethylene occurred from a high pressure pipe joint in the enclosed, unventilated ground floor area underneath a compressor house. The

Figure 2.5   Some of the damage caused by the explosion of a leak of ethylene in a confined space.

leak ignited and four men were killed. Figure 2.5 shows some of the damage. The source of ignition was never established with certainty but may have been faulty flameproof electrical equipment, as the investigators found that much of the equipment was badly maintained, with gaps too large and screws missing (see Section 4.3.1 on page 49). The joint that leaked was also badly made, and afterwards no expense was spared to improve the standard of joint-making; better tools, training and inspection were provided and the job was given to an elite group of fitters. This had been the practice in the past but a change had been made and all the fitters were trained to the required standard, or so it was thought. After the explosion the old system was restored.

Other recommendations made in the report on the explosion were similar to those made after the 1956 explosion:
• Surround the compressors and associated equipment by a steam curtain to prevent leaks reaching a source of ignition.
• Install flammable gas detectors so that leaks will be detected promptly.
• Install remotely operated valves so that leaking compressors can be isolated and blown down from a safe distance.
• Locate the compressors in an open-sided building so that small leaks can be dispersed by natural ventilation.

The last recommendation was not adopted immediately. The staff were concerned that if the compressors were installed in an open-sided compressor house the standard of maintenance would suffer. The rebuilt compressor house was enclosed, but a new plant, built a few years later after the conservative chief engineer had retired, was open-sided. There were no problems with maintenance.

Another company, operating similar plant, developed portable heaters and screens for use during winter maintenance[6].

The recommendations made after the two explosions were similar. The recommendations made after the first explosion were not passed on from one part of the company to another or, if they were passed on, their significance was not realised. If we handle ethylene as a gas it may not be immediately obvious that we can learn anything from a plant that handles propylene as a liquid. When we hear of an accident on another plant our first reaction is to think of all the reasons why our plant is different and the accident could not happen to us. We do not want to have to change our equipment or methods of operation.

Three of the men killed were maintenance workers who were repairing a compressor. The leak ignited several minutes after it was first noticed so they had ample time to leave. No-one told them to do so and no alarm was sounded as the leak was not treated as a hazard. No-one expected it to ignite. Another recommendation therefore was that leaks should trigger an alarm and all those who are not needed to deal with them should leave the area.

At the time of the second explosion the company had been using ethylene for thirty years. A memorandum issued when they started to use it said, 'In all the considerations that have been given in the past to this question, it has always been felt that excellent ventilation is a *sine qua non*. Not only does it reduce the danger of explosive concentrations of gas occurring, but it also protects the operators of the plant from the objectionable smell and soporific effects of ethylene'. During the following years this sound advice was forgotten or ignored, and it was not relearned when an explosion occurred in an associated company less than 20 km away. (See also Appendix 7.1, page 120.)

Reference 7 gives more details of the second explosion.

INCIDENT 3
A few years later an explosion occurred in a third company forming part of the same group. Again an unventilated compressor house was involved, this time one in which hydrogen was handled. A leak exploded and the roof and walls, both made of metal sheets, disappeared leaving only the steel framework to which they had been attached. The staff knew that there had been an earlier explosion in an ethylene compressor house, but they did not pay any attention

to the lessons learnt as hydrogen is much lighter than ethylene and disperses readily, or so they believed.

RECENT DEVELOPMENTS

As a result of these incidents and others[8], during the 1970s compressors and other equipment liable to leak were usually installed in open-sided buildings. In recent years closed buildings have again been built in order to meet new noise regulations. The buildings are usually provided with forced ventilation but this is much less effective than natural ventilation and is usually sized to provide the degree of ventilation necessary for the comfort of the operators rather than the dispersion of leaks. In New Zealand in 1986 I saw two plants which were outstanding by world standards. No expense had been spared to make sure that the most up-to-date safety features were installed, with one exception: both had completely enclosed compressor houses. The plants had to meet stringent noise regulations and enclosing the compressors was considered the best way of complying. One of the plants was out in the country; there was not a house in sight and the regulations seemed to have been made for the benefit of the sheep.

I saw a similar enclosed compressor house in the UK in 1991. The local authority had asked for a noise level of 35 dBA or less in the nearest habitations, a few isolated farms. Driving home after the visit I passed many rows of houses alongside the motorway where the noise levels were far higher, at a level which many people would find intolerable. Authorities are not always consistent.

Perhaps modern compressors leak less often than those built twenty or thirty years ago and there is less need for good ventilation. But perhaps another explosion is on the way.

We can reduce the noise radiation from compressors in other ways. For example, we can surround them with a close-fitting enclosure and purge the space in between continuously with air which is monitored for the presence of flammable gas.

A recent compressor explosion was the result of partially enclosing a compressor canopy, probably to aid winter operation and maintenance. A leak occurred because a spiral wound gasket had been replaced by a compressed asbestos fibre (caf) gasket seven years before. This was probably intended as a temporary measure but, once installed, the caf gasket was replaced by the same type during subsequent maintenance[9].

AN HISTORICAL NOTE

The willingness of companies to build in the open depends to some extent on their history.

Immigrants bring their customs with them. Nineteenth century German immigrants to South Australia built houses with steep roofs so the snow would slide off, though it never snows in South Australia. In the same way the oil companies started in hot countries and built their plants outdoors; they continued to do so when they moved to colder climes. Chemical companies started in Northern Europe, built indoors and were reluctant to change.

2.3 CHOKED VENTS

INCIDENT 1

In 1948 a vegetable oil was being hydrogenated in a pilot plant and samples had to be taken from a line joining two reactors (Figure 2.6). The sample pot was filled by opening the isolation valve and cracking the fine adjustment valve. Both valves were then closed and the sample pot drained into the sample can.

Two weeks after this system had been installed the operator was unable to get a sample and informed his chargehand. Suspecting a choke — the raw material and product both melted at 20°C — the chargehand tried to clear it by opening the isolation and drain valves, and then gradually opening the fine adjustment valve, so as to get a blow of gas straight through. The vent pipe suddenly moved and hit the chargehand on the head; he died later from his injuries. Hard hats were not normally worn at the time.

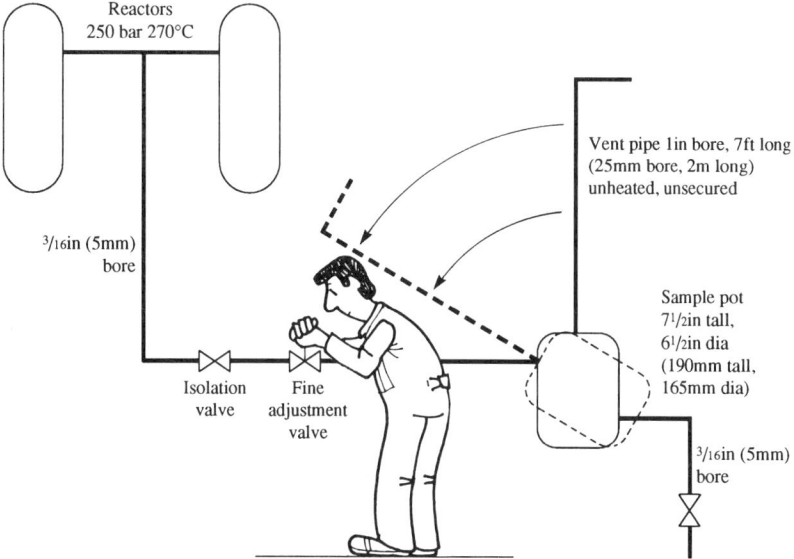

Figure 2.6   An attempt to clear a choke and obtain a sample from the line joining two reactors resulted in a fatal accident.

15

LESSONS FROM DISASTER

The sample pot and vent line were not clamped and were free to rotate by partially unscrewing the flange on the inlet pipe. It had been intended to clamp them but this had been overlooked. The sample line and pot were heated, but the vent line was not. It is believed that there was a choke somewhere in the system, probably in the unheated vent line, and when it cleared the sudden rush of gas caused the vent pipe to move backwards.

The main recommendation was that plants handling materials which are solid at ambient temperature should be adequately heated.

INCIDENT 2

In 1968, 200 m away on the same site, though in an area controlled by a different department, the end of a tank was blown off, killing two men who were working nearby on other equipment. The tank (Figure 2.7) was used for storing as liquid a product which melts at about 100°C. It was heated by a 7 bar gauge (100 psig) steam coil. The line into the tank was being blown by compressed air to prove that it was clear, the usual procedure before filling the tank. The vent on the tank, an open hole 3 inches (75 mm) in diameter, was choked and the air pressure (5.5 bar gauge or 80 psig) was sufficient to burst the tank (design pressure 0.35 bar gauge or 5 psig). The vent hole was not heated.

Everyone knew that the vent was liable to choke, but this was looked upon as an inconvenience rather than a danger. The operators, and their foremen and manager, did not realise that the air pressure was sufficient to burst the tank. Better access, so that it was easier to see if the vent was choked, and to rod it clear if it was, had been requested but got the priority given to inconveniences, not that given to safety measures. It was never provided.

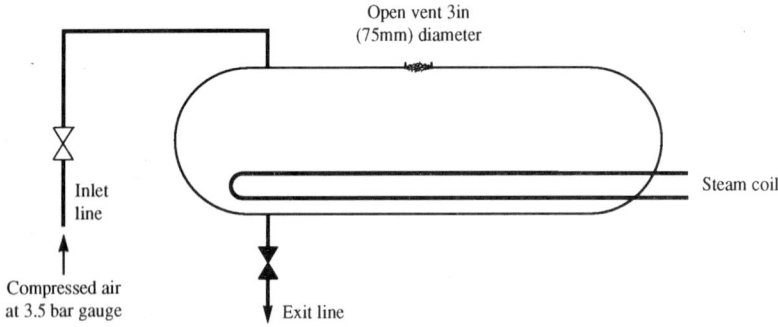

Figure 2.7   The unheated vent was choked and the air pressure was sufficient to blow the end off the tank.

ORGANISATIONS HAVE NO MEMORY

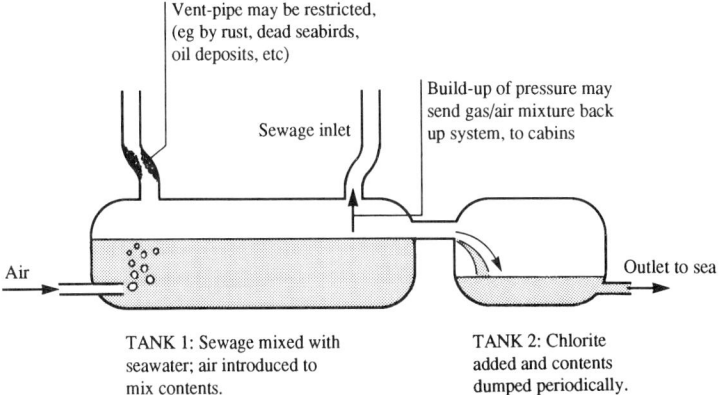

Figure 2.8(a)   Gas can flow back towards cabins from ship's sewage tank.

There were many other things wrong in this case, described in Reference 10. For example, the vent had originally been 6 inches diameter but had been reduced to 3 inches without authority.

The first incident was completely unknown to those who were working on the plant where the second incident occurred. Someone in the company safety department vaguely remembered it and after a search the original report was found.

INCIDENT 3 — ANOTHER INDUSTRY
In August 1992, while I was writing this book, a choked vent killed two children on a ferry travelling between Swansea in Wales and Cork in Ireland. Sewage from the toilets flowed by gravity into a vented tank where it was mixed with seawater and aerated with compressed air; the mixture then overflowed into a second tank where it was chlorinated before discharge to sea (Figure 2.8(a)). The vent and the usual toilet U-bends prevented gases from the first tank flowing backwards into the cabins.

The vent was choked and so compressed air and hydrogen sulphide from the sewage tank bubbled through the water in a U-bend, entered a cabin and killed two children who were asleep in their bunks. The air conditioning which would normally remove any gas from the cabin was switched off.

17

LESSONS FROM DISASTER

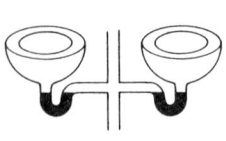  

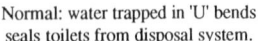

Normal: water trapped in 'U' bends   Ship tilts violently.   Toilets without water in 'U' bends
seals toilets from disposal system.   Water drains out of 'U' bends. are open to backflow of gas.

Figure 2.8(b) Gas can enter ship's cabins through drained U-bends.

The sea was rough and the tilting of the ship may have emptied some of the U-bends and provided an easier passage for the gases (Figure 2.8(b)).

A marine waste disposal engineer was quoted in the press[11] as saying that on this system, not the most modern, blockages of the vent pipes with grease, rust, dead birds and sanitary refuse, 'were not uncommon'. If so, the vents should have been checked regularly.

After the accident many passengers on earlier crossings said that they had complained about the smells in the cabins but their complaints were brushed aside or the cabins were merely sprayed with deodorant to mask the smells. Reluctance to recognise that there is a problem and that something should be done about it is a common failing (see Section 8.1, page 125). Another common failing is treating symptoms instead of causes, in this case treating the smell rather than its cause.

Another incident occurred on a preserved railway. The overflow pipe on a stainless steel water tank became blocked and while it was being filled the water pressure was sufficient to first balloon it and then split it[15].

Perhaps they will now check the overflow pipe at regular intervals. For how long?

2.4   PRESSURE VESSELS OPENED BY OPERATORS

Every day on every plant equipment which has been under pressure is opened for repair, normally under a permit-to-work system. One man prepares the equipment and issues a permit to another man who opens up the equipment, usually by carefully slackening the bolts in case any pressure is left inside. The involvement of two people and the issue of a permit provides an opportunity to

18

check that everything necessary has been done. Accidents are liable to happen when the same man prepares the equipment and opens it up, as the following incidents show.

INCIDENT 1

In 1965 a suspended catalyst was removed from a product in a pressure filter. After filtration was complete, the liquid left in the filter was blown out with steam. The pressure was then blown off through a vent valve and the fall in pressure watched on a pressure gauge (see Figure 2.9(a)). The operator then opened the filter for cleaning. The filter door was held closed by eight radial bars which fitted into U-bolts on the filter body (see Figure 2.9(b) on page 20). The bars were withdrawn from the U-bolts by turning a large wheel attached to the door. The door could then be opened.

One day an operator started to open the door before he had blown off the pressure. The door flew open with great violence. The operator was standing in front of it and was killed instantly.

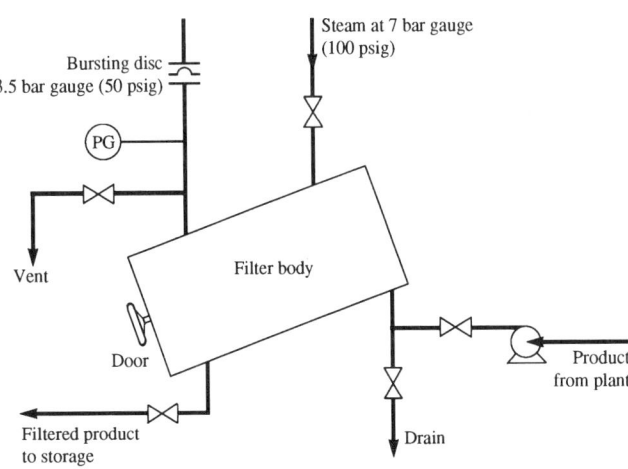

Figure 2.9(a) A pressure filter for removing suspended catalyst.

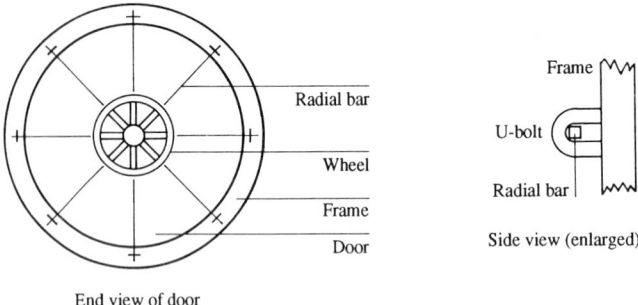

Figure 2.9(b) Filter door and detail of its fastening mechanism.

In this sort of situation it is almost inevitable than sooner or later the operator will forget that he has not blown off the pressure and will open the door before he has done so. To make repetition less likely, two changes were made to the filters:

• The operating wheel was modified so that the operator did not have to stand in front of the door.
• The pressure gauge and vent were moved nearer to the door so that the operator could see them when he was about to open the door.

Today we would not regard this action as sufficient. In addition. we would want two further features:

• An interlock so that the door cannot be opened until the vent valve is open.
• A two-stage opening mechanism: the first operation allows the door to open about 6 mm (¼ inch) while still capable of carrying the full pressure. If any pressure is present it can be allowed to blow off through the gap or the door can be resealed. A second operation is required to open the door fully.

The operators were surprised that a pressure of 'only 30 pounds' could cause the filter door to fly open with such violence. An explosions expert had to be brought in to convince them that there had not been a chemical explosion. The operators did not realise that a force of 30 pounds was exerted on every square inch of the door (area 10 square feet or 1 m^2), and that the total force was therefore 20 tons.

Reference 12 gives more details of this incident.

INCIDENT 2

Nine years later, in 1974, in the same company, plastic pellets were blown out of a road tanker by compressed air at a gauge pressure of 0.7 bar (10 psig). When the tanker seemed to be empty, the driver opened a manhole cover on the top and looked inside to make sure. One day the driver started to open the manhole cover before releasing the pressure. When he had opened two of the quick-release retainers, the cover was blown open and the driver was blown off the tanker and killed by the fall.

The driver's error may have been due to a moment's forgetfulness, something that happens to us all from time to time. Or, as he was not the regular driver, he may have thought that he could let the pressure blow off through the manhole.

After the accident the manhole cover was replaced by one with a two-stage opening device, as described above, and the vent valve was moved from the side of the tanker to the foot of the ladder. The engineering department specifications were changed to state that two-stage mechanisms are required on all pressure vessels opened by operators.

As in the first incident the dead man's colleagues were surprised that 'a puff of air' could blow a man off the top of a tanker and they thought there must have been a chemical explosion. However, 0.7 bar (10 psig) is not a small pressure. An explosion which develops this pressure can level a plant.

INCIDENT 3

A few years later, in the same company, two operators opened a large filter attached to a 14 inch (0.35 m) liquid propane line on a new plant before the pressure had been blown off. Fortunately the leak did not ignite, but the operators were hospitalised with cold burns. A two-stage opening mechanism had not been fitted as the filter was classified as a pipe fitting and designed by the piping section of the engineering department who were not familiar with the vessel specifications.

2.5   IMPROVING THE CORPORATE MEMORY

How can we prevent serious accidents such as those I have described repeating themselves after ten or more years? Here are some suggestions. They are discussed further in Chapter 10.

- In every instruction, code and standard make a note on the reason why. Add accounts of accidents which would not have occurred if the instruction, code or standard had been followed.

- Describe old accidents as well as recent ones in safety bulletins and newsletters and discuss them at safety meetings. 'Giving the message once is not enough.'[13]

- Follow up at regular intervals to see that the recommendations made after accidents are being followed, in design as well as operations. The recommendation made after the first incident — that equipment under repair should be isolated by slip-plates — may have lapsed because the engineering department continued to design plants that could not easily be slip-plated.

- Remember that the first step down the road to the next accident in this series occurs when someone turns a blind eye to a missing blind.

- Never remove equipment before you know why it was installed. Never abandon a procedure before you know why it was adopted.

- Devise better retrieval systems so that we can find, more easily than at present, details of past accidents, in our own and other companies, and the recommendations made afterwards. Few companies make full use of the computerised information systems that are available and even where there is a system, few employees seem to use it. If such a system is to be widely used it must available at every desk. It will not be widely used if we have to walk to the library and/or consult an information scientist.

- Include important accidents of the past in the training of undergraduates and company employees. Compared with the large number of courses on hazard identification and assessment, few courses cover past accidents although suitable training material is available from the Institution of Chemical Engineers, UK. Sets of slides and notes[14] illustrate accidents that have occurred and the action needed to prevent them happening again. Each set covers a topic such as Preparation for Maintenance, Hazards of Plant Modifications or Human Error. They can be used as lecture material but are more effective if the group discusses the reasons for the accidents and says what *they think* should be done to prevent the accidents happening again. Such discussions will have more impact if the group discusses local accidents, but details and photographs are often lacking.

Employees at all levels, from operators to senior managers, should take part in the discussions, which should be repeated from time to time. I consider these discussions further in Section 10.2 on page 167.

REFERENCES IN CHAPTER 2
1. Nightingale, P.J., January 1989, *Plant/Operations Progress*, 8 (1): 28.
2. *The Phillips 66 Company Houston Chemical Complex Explosion and Fire*, April 1990 (US Dept of Labor, Washington, DC, USA).
3. Cullen, W.D., 1990, *The Public Inquiry into the Piper Alpha Disaster* (HMSO, London, UK).

4. *Health and Safety at Work*, March 1992, 4 (3): 5.
5. Stratton, W.E., 1989, *A Review of Criticality Incidents*, Report No DOE/NCT–04 (US Dept of Energy, Washington, DC, USA).
6. Morris, D.H.A., 1974, in *Loss Prevention and Safety Promotion in the Process Industries*, edited by C.H. Buschmann (Elsevier, Amsterdam, The Netherlands), 369.
7. Kletz, T.A., 1988, *Learning from Accidents in Industry* (Butterworths, London, UK), Chapter 4.
8. Howard, W.B., 1972, *Loss Prevention*, 6: 58.
9. MacDiarmid, J.A. and North, G.J.T., 1989, *Plant/Operations Progress*, 8 (2): 96.
10. Kletz, T.A., 1988, *Learning from Accidents in Industry* (Butterworths, London, UK), Chapter 7.
11. *Daily Telegraph*, 14 August 1992, 2.
12. Kletz, T.A., 1991, *An Engineer's View of Human Error*, 2nd edition (Institution of Chemical Engineers, Rugby, UK), Section 2.2.
13. Davy, Sir Humphrey, writing in 1825. Quoted by McQuaid, J., 1991, *Process Safety and Environmental Protection*, 69 (B1): 9.
14. *Hazard Workshop Modules*, Collections of notes and slides on accidents that have occurred and the action needed to prevent them happening again (Institution of Chemical Engineers, Rugby, UK), various dates.
15. *Ffestiniog Railway Magazine*, Autumn 1992, No 138, 221.

APPENDIX 2.1

'IT MUST NEVER HAPPEN AGAIN'
A sketch for chemical engineers in two Acts

SCENE:
A committee room.

ACT I

TIME:
Any time during the 20th century.

Enter **Chairman** and three Directors (**A**, **B** and **C**).

Chairman: The first item on our agenda is the report on last month's explosion.

A: Well, as you know, the incident was due to the explosion of a crapgrinder and it occurred as the result of errors in design and methods of operation which were not picked up by our existing procedures. It is fortunate that no-one was killed, but there has been extensive damage to plant which will take us a long time to repair and there has been a serious loss of production.

    I support the recommendation of the investigating committee that a special group of people should be appointed to devote themselves full-time to the design and examination of crapgrinders, writing operating procedures and checking up to see that they are followed. They will keep in touch with developments in crapgrinder technology elsewhere in the world.

    A team of five men will be needed. I am hoping to arrange for Dr Nitpicker to be released from his present job so that he can head up this team.

B: Can we afford to release him? He was earmarked as project engineer of our new joint venture.

Chairman: We shall have to. We cannot afford another incident like the one we have just had. Quite apart from the damage to plant and loss of production, it has badly damaged our reputation with the public and we must show that we are determined never to let anything like this happen again.

A: What is more, we have found to our annoyance that the *Crapgrinder Regulations 1882* are still in force. We are in trouble with the Health and Safety Executive because we did not report the incident and we have had their wretched inspectors wandering all over the plant.

APPENDIX 2.1

C: We are never going to recover our lost share of the market. Five salaries are a minute proportion of the sums that are at stake. I fully support the proposal.

B: Why can't the Works undertake the necessary inspections?

A: The existing staff do not have the time or the expertise to carry them out with the thoroughness and detail that is required.

C: Nor do they have the time to keep in touch with new developments worldwide.

Chairman: The technology is too specialised for the ordinary plant engineer.

C: Inspectors must be independent of the production team or they will be swayed by production pressures.

Chairman: Who else will be in the team besides Dr Nitpicker?

A: We haven't yet decided, but there will be at least two or three senior engineers covering a range of experience.

Chairman: The report is accepted. Dr Nitpicker must be appointed without delay whatever the inconvenience to other departments. I want to see at least some of his team appointed by the end of the month. Whatever happens, IT MUST NEVER HAPPEN AGAIN!

ACT II

TIME:
Ten years later.

Enter a different **Chairman** and two Directors (**D** and **E**).

Chairman: We'll start today with the Personnel Report.

D: Dr Nitpicker has agreed to accept early retirement and he will not be replaced, so there will be a reduction of one in the senior staff.

E: What did he do?

D: Oh, he had some sort of inspection or advisory job in the production area. Not sure what. I think he was an expert in crapgrinders or something equally specialised.

Chairman: We cannot afford the luxury of employing large numbers of people in advisory and inspection functions. Our competitors don't do it. It is the responsibility of the Works to inspect their equipment and to see that they have sufficient knowledge of the way it works.

E: Wasn't there an explosion once on a crapgrinder?

D: I believe there was, but a long time ago — a generation or more ago. I expect designs have changed since then. And just because a piece of equipment blows up once, there is no need to employ specialists on it forever afterwards. If we employ specialists on crapgrinders, somebody will say that we should have specialists on piping or machines or vessels, and that would be absurd. We are trying to do away with most of these narrow-minded specialists. All they do is create work for other people. Once we've got rid of Dr Nitpicker he won't be able to tell us about problems and we don't have to do anything about them.

Chairman: We need specialists in subjects like Personnel, but any competent engineer should be familiar with the whole range of equipment under his control. We have agreed to the appointment of an extra Assistant Personnel Manager so we must economise in staff elsewhere.

E: Who will supervise Dr Nitpicker's staff?

D: Oh, he hasn't had any for many years. He has been a lone worker. That makes it so easy to get rid of him. We don't have the problem of deciding what to do with his staff.

Chairman: Right. It's agreed then. Let's move on to the next item.

An explosion is heard off-stage.

Voice (*off-stage*): The crapgrinder's exploded!

Curtain

3. ACCIDENTS OF THE COMING YEAR

'"It's a poor sort of memory that only works backwards", the Queen remarked. "What sort of things do you remember best?", Alice ventured to ask. "Oh, the things that happened the week after next", the Queen replied in a careless tone.'
Lewis Carroll, *Through the Looking Glass*

'He only reads eighteenth-century newspapers of which he has an enormous stock, for he says the news in them is just the same as it is today. You merely have to substitute the names of countries occasionally, and not invariably.'
Professor Sir Albert Richardson, described in *National Trust*, Summer 1975, No 23, 13.

This chapter describes a number of accidents and incidents which will occur in most large and many small chemical companies during the coming year. They are not major accidents, such as those described in Chapter 2, but they will cause localised damage and loss of production. Some of them will cause injury and could possibly escalate to more serious proportions.

Why am I so confident that they will occur? I am not *certain* that they will occur, just as I am not certain that the sun will rise tomorrow morning. I expect that it will, as it has done so regularly in the past and I can see no change in circumstances that may prevent it rising in the future. In the same way, the accidents I shall describe have happened frequently in the past and there is little sign of sufficient change to prevent them occurring in the future.

However, although I can forecast with reasonable confidence what will happen, I cannot, unfortunately, say where and when it will occur.

3.1 A TANK WILL BE SUCKED IN

The collapse of atmospheric pressure storage tanks has been a frequent occurrence in almost every company[1,2] and more tanks will be sucked in during the coming year. The precise method is uncertain, but here are some of the ways in which tanks have been collapsed in the past. In most cases the triggering event, not the cause, was the withdrawal of liquid from the tank by pumping or draining.

- The flame arrestor in the vent pipe was not cleaned for over two years, though

LESSONS FROM DISASTER

Figure 3.1 This tank was sucked in while it was being emptied, because the flame arrestors in the three vents had not been cleaned for over two years.

it was scheduled for cleaning every three months (Figure 3.1) (see Section 10.1, page 166).

• A loose blank was put on top of a vent to stop fumes coming out near a walkway.

• A plastic sheet was tied over a vent to stop dust getting in while the pressure/vacuum valve was being overhauled.

• A thunderstorm occurred while a tank was being steamed and the vent was not big enough to prevent it being sucked in when the steam condensed.

• A pressure/vacuum valve was corroded by the material in the tank.

• A larger pump was connected to a tank and it was pumped out more quickly than air could get in through the vent.

• Before emptying a road tanker the driver propped the manhole lid open. It fell shut.

• Water was added too quickly to a tank containing a solution of ammonia in water.

• A tank was fitted with a pressure relief valve but no vacuum relief valve. When cold liquid was added to hot liquid already in the tank the pressure fell.

• A tank was boxed up with some water inside and rust formation used up the oxygen in the air.

- To prevent fumes coming out of a vent while men were working nearby, a hose was connected to the vent and the other end put in a drum of water. When the tank was emptied water rose up the hose and then the tank was sucked in.
- Tank A was vented through tank B (Figure 3.2). Tank B was slip-plated for entry. When liquid was pumped out of tank A it collapsed.
- A vent was almost blocked by polymer. The liquid in the tank was inhibited to prevent polymerisation but the vapour that condensed in the vent was not inhibited.

There are no prizes for finding new methods of sucking in tanks or for sucking in more tanks, by more methods, than other plants.

The common feature in all these incidents is that operators, and sometimes supervisors, did not understand that large storage tanks are very fragile. They are usually designed to withstand a vacuum of only 2½ inches (65 mm) water gauge (0.6 kPa), the pressure at the bottom of a cup of tea. Most people can blow or suck a pressure at least 20 times higher. If a tank was airtight we could suck it in without any difficulty, though it would take us a long time. A training handout on the strength of storage tanks is reproduced in the Appendix to this chapter; it is also available in the form of slides[3].

3.2 A TRIP WILL FAIL TO OPERATE

Many trips which are intended to close or open a valve when a temperature, pressure, level or concentration gets too low or too high will fail to operate; as

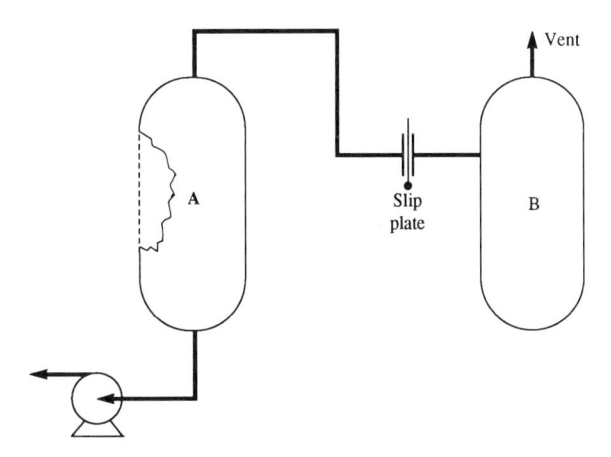

Figure 3.2 Tank B was slip-plated for entry. When liquid was pumped out of tank A, it collapsed.

a result tanks will overflow, equipment will get too hot, reactions will run away and many similar incidents will occur. The trips will fail for many reasons:

THE TRIP IS NOT TESTED REGULARLY

All trips develop fail-danger faults (that is, faults which prevent them operating when required) from time to time, and these failures can be detected only by testing. The test frequency should depend on the failure rate (typically once every few years) and the demand rate, that is, the frequency with which the trip is expected to operate. Typical test frequencies are once a month or once in three months[4]. Sometimes there is no regular test schedule; sometimes the scheduled tests are too infrequent; sometimes the schedule is not followed.

THE TEST IS NOT THOROUGH

A trip has three parts: the measuring device or sensor, the processing unit (which can be a microprocessor or conventional hard-wired equipment) and the valve. All three should be tested. The sensor should be tested by changing the pressure, level or concentration to which it is exposed. (If temperature is being measured a current may be injected from a potentiometer.) The valve should, if possible, be tested by closing (or opening) it fully. If this upsets production it may be closed halfway and tested fully during a shutdown. Sometimes, however, the valve is not tested; sometimes only the processing unit is tested; sometimes the sensor is tested but the level at which it operates or its speed of response is not checked.

THE TEST IS NOT LIKE 'REAL LIFE'

For example, a trip should not be tested by raising (or lowering) the setpoint as this does not test the ability of the sensor to reach the normal setpoint.

A trip was removed from its case before testing. When it was required to operate in anger, a pointer fouled the case and could not move freely.

THE SETPOINT HAS BEEN ALTERED

Set points should be changed only after authorisation in writing at an agreed level of management.

THE TRIP HAS BEEN MADE INOPERATIVE (DISARMED)

Sometimes this is done because it operates when it should not and upsets production; sometimes it is done so that it can be tested or repaired or because a low flow or low temperature trip has to be isolated during start-up.

A turbine on a North Sea oil platform was routinely cleaned with high pressure water. Chemicals in the water affected the flammable gas alarms, so they were isolated. When cleaning was complete the alarms were not reset. When a flameout occurred it was not detected, gas continued to build up in the combustion chamber and an explosion occurred, followed by a fire[5].

Managers and supervisors should carry out regular checks. If a by-passed trip causes an accident, it has probably been by-passed for some time or on many occasions, and an alert manager or supervisor could have spotted it. If a trip has to be isolated for start-up, the isolator can be fitted with a timer so that the trip is automatically re-armed after an agreed period of time.

RANDOM FAILURE

Finally, even if the trip is tested throroughly and regularly, it may still fail at random (due to an internal fault) after the last test and before it is required to operate. This is the least likely cause of failure.

Even if a company has a good test programme, standards may not be as high in some of the smaller, out-of-the-way plants. There is more on trip testing in Sections 6.2 (page 74), 7.5 (page 101) and 8.8 (page 134).

3.3 A ROAD TANKER WILL BE OVERFILLED

We can be sure that this will occur, perhaps many times, during the coming year.

Road tankers that are filled slowly are overfilled because the operator leaves the job and is delayed or, for some reason, filling is quicker than usual. When he returns the tanker is overflowing.

Most overfilling incidents occur, however, on installations where filling is fast and is controlled by meters. Sometimes the meter is faulty, sometimes it is set incorrectly, sometimes there is a residue in the tanker from the previous load. Many companies now fit high level trips on every compartment of every tanker. Like all trips (see Section 3.2, page 29) they should be tested regularly.

Tankers have been overfilled because a new shift did not know that the outgoing shift had already filled a tanker and tried to fill it again.

Even though the liquid spilt is flammable, in most cases it does not ignite. But sometimes it does and there have been some spectacular fires[6].

3.4 A MAN WILL BE INJURED WHILE DISCONNECTING A HOSE

Someone will disconnect a hose while it is under pressure or full of liquid (as there is no other way of relieving the pressure or emptying the hose) and will be splashed with process material or injured by a sudden movement of the hose.

Afterwards there will be a campaign to fit vent valves on all hoses or to all points to which they are likely to be connected. After a few years the incident will have been forgotten, hoses will again be in use without vents and the accident will recur.

Other accidents will also involve hoses:

- The wrong sort will be used.
- A damaged hose will be used.
- Different screw threads will be combined, the hose will be held by only a couple of threads and will come off during use.
- A steam, air or nitrogen valve at one end of a hose will be closed before the process valve at the other end; process liquid will then enter the hose which is not designed to withstand it[7].

3.5 A ROAD OR RAIL TANKER WILL BE MOVED BEFORE THE HOSE HAS BEEN DISCONNECTED

If filling or emptying is complete, the hose will be damaged. If filling or emptying is still in progress there will be a spillage.

A few incidents have occurred because vehicles were parked on a slope and their brakes failed or because another vehicle collided with them. But most incidents are due to a misunderstanding between the driver and the filler. In the UK many such incidents have occurred on the railways, with engines entering a siding and removing a train before filling was complete[8].

There are several ways for preventing such accidents or minimising their consequences:

- Lock the tanker inside the filling (or emptying) bay. Like all procedures this will lapse after a few months or years unless managers make a continuing effort to see that it is enforced. It is, of course, possible to interlock the gate and the filling valve.
- Fix a small lever to the road tanker in front of the point to which the hose is attached. To connect the hose this lever has to be moved aside. This locks the tanker's brakes and they cannot be released until the hose has been disconnected and the lever moved back to its original position.
- Fit a device (available commercially) inside the hose; it closes automatically if the hose is broken.
- Use remotely-operated emergency isolation valves which allow filling and emptying lines to be isolated rapidly from a safe distance if a hose fails.

- Finally, what will happen to the spillage? Is the ground sloped so that spillages flow into a drain or catchment pit? Remember that the level indicators on some tankers will not read correctly if the tanker is not level.

3.6 THE WRONG PIPELINE WILL BE OPENED

During the coming year someone will be given a permit to break into a pipeline, or weld a branch on a pipeline, which has been prepared for maintenance. The line will be shown to him, he will go for his tools and then he will break into the wrong line. Or he will know the plant so well that he does not need to be shown the line. Or the line will be marked with chalk but the chalk will be washed off by rain or he will go to the wrong chalk mark (Figure 3.3).

To prevent such accidents a numbered tag should be tied to the pipeline at the point at which it should be broken and the number of the tag should be written on the permit-to-work. Alternatively, a duplicate tag can be given to the maintenance worker (see Section 7.8 on page 109).

Figure 3.3 Which joint should be broken? A joint to be broken was marked with chalk. The fitter broke another joint which had older chalk marks on it.

LESSONS FROM DISASTER

Figure 3.4 This tank contained hot oil, above 100°C. When some water was accidentally added, the roof lifted and a structure 25 m tall was covered in black oil.

3.7   A HEAVY OIL TANK WILL FOAM OVER

During the coming year someone will add hot oil, at a temperature of over 100°C, to a heavy oil tank containing a layer of water. The water will vaporise with explosive violence and, if the operator is lucky and the vent is large, a mixture of steam and hot oil will be discharged through it. This is called foam over. If the operator is not so lucky, the roof will be lifted. In incidents of this sort structures 25 m (80 feet) tall have been covered with black oil. Figure 3.4 shows a tank ruptured by foam over.

To prevent such incidents the incoming oil should be cooled below 100°C and a high temperature alarm should be fitted to the oil line. If this is not possible, because the oil would then be too viscous, then the liquid in the tank should be kept above 100°C so that any water which gets into the tank is vaporised. In addition, before the hot oil is added any water present should be drained and the tank should be circulated.

3.8 A PIPE WILL BE DAMAGED BY WATER HAMMER

Such damage often occurs when steam mains, which have been shut down, are brought back on line. Much condensate is formed and the normal steam traps may not be able to remove it, especially if they have not been maintained in first class condition. If the condensate is collected for re-use, the backpressure in the condensate line may prevent the traps functioning when the pressure in the steam main is still low. By-passes to drain should be provided.

High pressure steam mains can be damaged by water hammer as well as low pressure mains. Figure 3.5 shows a 40 bar gauge (600 psig) steam main which ruptured, seriously injuring several men. All steam mains should be brought on line slowly and the condensate drains checked to make sure they are in working order.

Figure 3.5   This 40 bar steam main was broken by water hammer.

3.9 MUST THESE INCIDENTS OCCUR AGAIN?

Like the incidents described in Chapter 2, why do the simpler incidents described in this chapter continue to happen? Preventing them is not a difficult or insoluble technical problem. They occur because we do not use the knowledge we already have.

All higher animals learn by experience. Are chemical engineers and chemical plant operators exceptions? The answer is 'No'. Each individual learns by his experience, and it is unusual for the same accident to occur to the same person twice, but we are slow to learn from the experience of others. The suggestions made at the end of Chapter 2 and in Chapter 10 apply equally to the incidents discussed in this chapter.

'Modern man', writes the historian E.H. Carr, 'peers eagerly back into the twilight out of which he has come, in the hope that its faint beams will illuminate the obscurity into which he is going ...'[9] I wish it were so. For those who are willing to look, searchlights — not faint beams — shine out of the past and show us the pits into which we will fall if we do not look where we are going.

REFERENCES IN CHAPTER 3

1. Kletz, T.A., 1988, *What Went Wrong? — Case Histories of Process Plant Disasters*, 2nd edition (Gulf, Houston, Texas, USA), Section 5.3.
2. Sanders, R.E., Haines, D.L. and Wood, J.H., January 1990, *Plant/Operations Progress*, 9 (1): 63.
3. *Hazards of Over- and Under-Pressuring Vessels*, Hazard Workshop Module No 001 (Institution of Chemical Engineers, Rugby, UK), undated.
4. Kletz, T.A., 1992, *Hazop and Hazan — Identifying and Assessing Process Industry Hazards*, 3rd edition (Institution of Chemical Engineers, Rugby, UK; published in the USA by Taylor & Francis, Bristol, PA), Section 3.3.
5. *The Chemical Engineer*, 11 June 1992, No 521, 8.
6. Kletz, T.A., 1988, *What Went Wrong? — Case Histories of Process Plant Disasters*, 2nd edition (Gulf, Houston, Texas, USA), Section 13.1.
7. Kletz, T.A., 1990, *Critical Aspects of Safety and Loss Prevention* (Butterworths, London, UK), 163.
8. *Petroleum Review*, May 1974, 345.
9. Carr, E.H., 1990, *What is History?*, 2nd edition (Penguin Books, London, UK), 134.

APPENDIX 3.1 — HOW STRONG IS A STORAGE TANK?

A STORAGE TANK IS DESIGNED:

1. TO HOLD LIQUID

Liquid exerts pressure on the sides and base of the tank.

   Pressure = head of liquid.

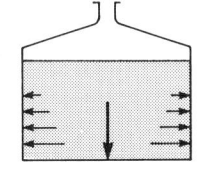

2. TO BE FILLED

For liquid to get in air and vapour must get out. If they can't the tank will be pressurised. For air and vapour to be pushed out the pressure in the tank must be slightly above atmospheric pressure.

   The tank is designed for an internal pressure of 8 ins water gauge (W.G.).

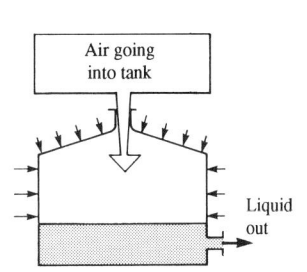

3. TO BE EMPTIED

For liquid to get out air must get in. If it can't the tank will be under-pressured. For air to be sucked in the pressure in the tank must be slightly below atmospheric pressure.

   The tank is designed for an external pressure (or vacuum in the tank) of 2½ ins W.G.

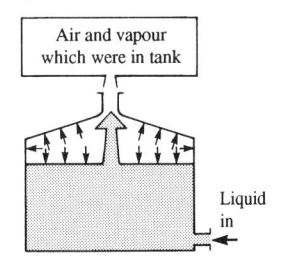

WHAT ARE INCHES WATER GAUGE?

They are a measurement of pressure, used for very low pressures:

8 ins (200 mm) W.G. = ⅓ psi (2.0 kPa)

2½ ins (65 mm) W.G. = ¹⁄₁₀ psi (0.6 kPa)

Or put another way:

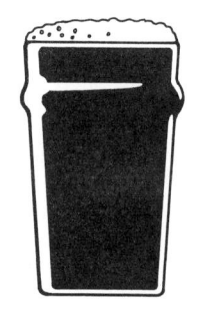

2½ ins W.G. is the pressure at the bottom of a cup of tea.

8 ins W.G. is the pressure at the bottom of a pint of beer.

YOU CAN BLOW OR SUCK AT LEAST 24 INS (600 mm) W.G.

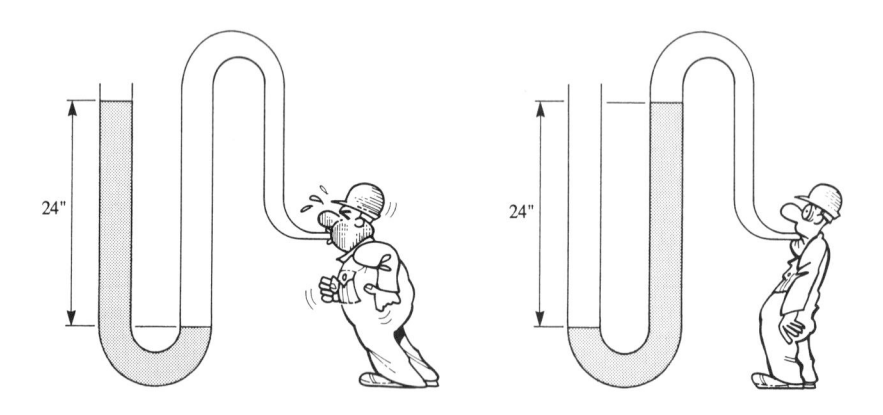

That means by just using your lungs you could over- or under-pressure a storage tank. (Because of the volume of air it would take you a long time.)

If you don't believe it, because storage tanks always look big and strong, just study the table on the next page.

APPENDIX 3.1

If a baked bean tin has a strength of 1, then:

	Shell	**Roof**
Baked bean tin (small)	1	1
40 gallon drum	½	⅓
50 m³ tank	⅓	⅛
100 m³ tank	¼	1/11
500 m³ tank	⅙	1/33
1000 m³ tank	⅛	1/57

Next time you eat baked beans just see how easy it is to push the sides or top in with your fingers — *and then look at the table again.* (Any small tin will do if you don't like baked beans.)

Note also:
- the bigger the tank the more fragile it is;
- the roof is weaker than the shell.

Up to 1000 m³ the tank shell and roof are only as thick as this:

IF ALL THAT'S TRUE — IS A STORAGE TANK STRONG ENOUGH?
Yes. A 1000 m³ tank has a factor of safety of 2 against failure provided it is operated within the very low pressure allowed. (Smaller tanks have bigger safety factors.)

Most of the pressures we have available are many times bigger than the allowable pressures, that is, 8 ins W.G. inside, 2½ ins W.G. outside.

For example:
Full atmospheric pressure outside	=	150 times bigger
Transfer pump head inside	=	120 times bigger
40 psi (2.75 bar) nitrogen inside	=	120 times bigger
100 psi (7 bar) steam inside	=	300 times bigger

All of these pressures or even a small part of them will cause the tank to:

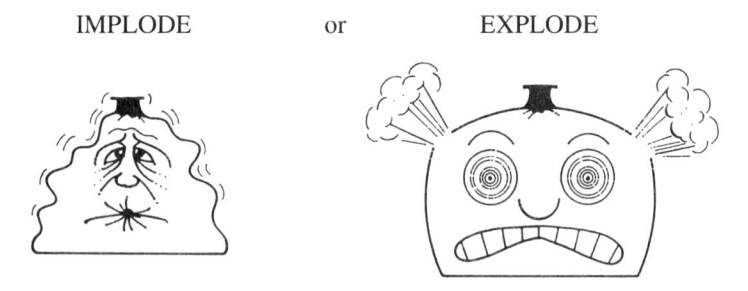

IMPLODE　　　or　　　EXPLODE

HOW DO WE STOP THIS HAPPENING?

By making sure that:

1. The tank has a vent big enough to relieve all sources of pressure that might be applied to it.
2. The vent is always clear.
3. The vent is never modified without the authorisation of the plant or section engineer.

Here are some typical faults in vents which should never happen:

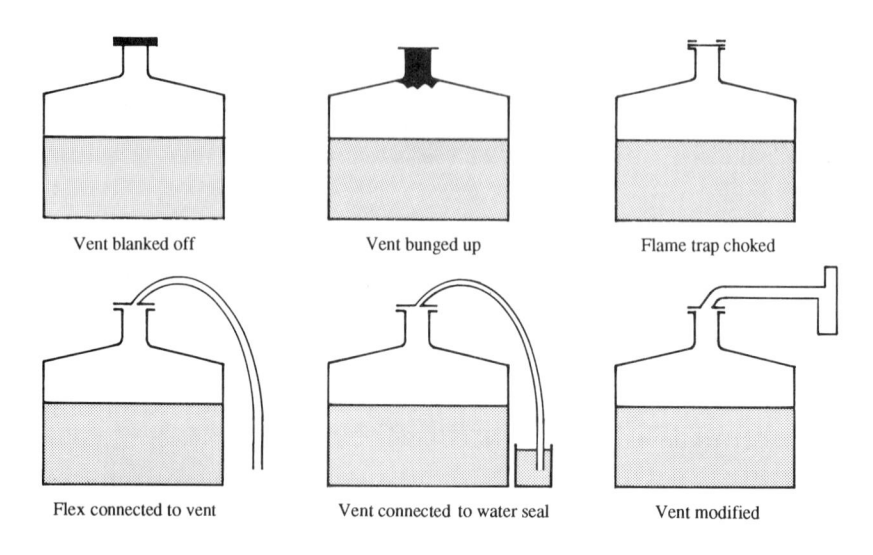

Vent blanked off　　　Vent bunged up　　　Flame trap choked

Flex connected to vent　　　Vent connected to water seal　　　Vent modified

Don't look surprised — one (or more?) of these has almost certainly happened on your plant in the last year — it could have had serious consequences.

KNOWLEDGE AND VIGILANCE CAN STOP IT HAPPENING AGAIN

4. SOME OTHER FAILURES TO LEARN FROM THE PAST

'Disasters ... happen when the decisions are made by people who cannot remember what happened last time.'
City comment, *Daily Telegraph*, 17 May 1990

4.1 KNOWLEDGE IN THE WRONG PLACE

Accidents have occurred because those concerned did not know how to prevent them although the knowledge was well known to other people. The knowledge was in the wrong place.

4.1.1 ABBEYSTEAD

In 1984 an explosion in a water pumping station at Abbeystead, Lancashire, killed 16 people, most of them local people who were visiting the site.

Water was pumped from one river to another. When pumping stopped some of the water was allowed to drain out of the tunnel, leaving a void which filled with methane from the rocks below. When pumping was restarted the methane was pushed out through vent valves which discharged inside an underground pumphouse, where it exploded.

If the operating staff had known that methane might have been present, they could have prevented the explosion by keeping the tunnel full of water (as the designers intended) or by venting the gas from the vent valves into the open air. In addition they could have prohibited smoking — the probable source of ignition — in the pumphouse, though they should not have relied on that alone; as we saw in Section 2.2 (page 10), flammable mixtures are easily ignited. None of these things were done because they did not realise that methane might be present. The official report[1] said, 'References to the presence of dissolved methane in water supply systems have been traced in published literature but they do not appear to have achieved wide circulation particularly amongst the sections of the civil engineering profession concerned with water supply schemes'.

Nevertheless, it is surprising that the vent discharged into the pumphouse instead of the open air. It seems that the local planning authority objected in principle to any equipment that might spoil the view.

The High Court decided that the designers, the operators and the builders were all responsible — in varying degrees — for the explosion and should share the cost of compensating the victims, but the Court of Appeal decided that the designers were wholly responsible[2].

The official report recommended that the hazards of methane in water supplies should be more widely known but this is just one piece of knowledge that needs to be more widely publicised. Making it better known may prevent a repetition of the last accident but will not prevent the next one. What can we do about the general problem? Perhaps engineers should be encouraged to read widely and acquire a ragbag of bits and pieces of knowledge that might come in useful at some time. If one member of the design or operating teams had known, however vaguely, that methane might be present and had said so, the explosion might not have occurred. A hazard and operability study can help us dredge up half-forgotten knowledge from the depths of our minds. Also, if we think there might be a hazard we should always speak up even though we are not responsible for that part of the plant or that aspect of design.

Another reason why engineers should read widely is that:

'Breadth of knowledge and a catholic interest are ... inherent in most discovery, invention and innovation ...

'... most discoveries or inventions occur when, in the mind of the scientist or engineer involved, two or more pieces of information (frequently from different sources) come together and some relationship or correspondence is acknowledged or suspected.'[3]

Never assume that the experts know everything they ought to know. Never say, 'They might be offended if I tell them how to do their job. I expect they know what they are doing. Better say nothing'.

The following incidents support this and my other suggestions.

4.1.2 FLIXBOROUGH

The explosion at Flixborough, Humberside, in 1974 is well known. A temporary pipe replaced a reactor which had been removed for repair. The pipe was not properly designed (designed is hardly the word as the only drawing was a chalk sketch on the workshop floor) and was not properly supported; it merely rested on scaffolding. The pipe failed, releasing about 30–50 tonnes of hot hydrocarbons which vaporised and exploded, devastating the site and killing 28 people[4-6].

The reactor was removed because it developed a crack and the reason for the crack illustrates the theme of this section. The stirrer gland on the top of the reactor was leaking and, to condense the leak, cold water was poured over the top of the reactor. Plant cooling water was used as it was conveniently available. Unfortunately it contained nitrate which caused stress corrosion cracking of the mild steel reactor (which was lined with stainless steel).

Afterwards it was said that the cracking of mild steel when exposed to nitrates was well known to materials scientists but it was not well known — in fact hardly known at all — to chemical engineers, the people in charge of plant operation.

The temporary pipe and its supports were badly designed because there was no professionally qualified mechanical engineer on site at the time. The works engineer had left, his replacement had not arrived and the men asked to make the pipe had great practical experience and drive but did not know that the design of large pipes operating at high temperatures and pressures (150°C and 10 bar gauge [150 psig]) was a job for experts. There were, however, many chemical engineers on site and the pipe was in use for three months before failure occurred. If any of the chemical engineers had doubts about the integrity of the pipe they said nothing. Perhaps they felt that the men who built the pipe would resent interference. Flixborough shows that if we have doubts we should always speak up.

4.1.3 ABERFAN

At Aberfan in South Wales in 1966 a colliery tip collapsed onto the village, killing 144 people, most of them children. The immediate technical cause was simple: the tip was placed over a stream on sloping ground, but the official report[7,8] brought out the underlying causes very clearly. They included a failure to learn the lessons of the past, a failure to inspect regularly and a failure to employ competent people with the right professional qualifications. To quote from the report:

'... forty years before it occurred, we have the basic cause of the Aberfan disaster recognised and warned against. But, as we shall see, it was a warning which went largely unheeded. (Paragraph 44.)

'Tip slides are not new phenomena. Although not frequent, they have happened throughout the world and particularly in South Wales for many years, and they have given rise to an extensive body of literature available long before the disaster. (Paragraph 72.)'

In 1939 there was a tip slide at Cilfynydd in South Wales.

'Its speed and destructive effect were comparable with the disaster at Aberfan, but fortunately no school, house or person lay in its path. ... It could not fail to have alerted the minds of all reasonably prudent personnel employed in the industry of the dangers lurking in coal-tips ... the lesson if ever learnt, was soon forgotten. (Paragraph 82.)

'In 1944 another tip at Aberfan slid 500–600 feet. Apparently, no-one

LESSONS FROM DISASTER

troubled to investigate why it had slipped, but a completely adequate explanation was to hand ... it covered 400 feet of a totally unculverted stream. (Paragraph 88.) To all who had eyes to see, it provided a constant and vivid reminder (if any were needed) that tips built on slopes can and do slip and, once having started, can and do travel long distances. (Paragraph 89.)

'Why was there this general neglect? Human nature being what it is, we think the answer to this question lies in the fact that ... there is no previous case of loss of life due to tip stability ... (Paragraph 68.)'

It was customary in the coal industry for tips to be the responsibility of mechanical rather than civil engineers.

'It was left to the mechanical engineer to do with tips what seemed best to him in such time as was available after attending to his multifarious other duties. (Paragraph 70.) For our part, we are firmly of the opinion that, had a competent civil engineer examined Tip 7 in May, 1965, the inevitable result would have been that all tipping there would have been stopped. (Paragraph 73.) The tragedy is that no civil engineer ever examined it and that vast quantities of refuse continued to be tipped there for another 18 months. (Paragraph 168.)'

4.1.4 A PIPELINE FAILURE

Figure 4.1 shows part of a cross-country ethylene pipeline. The two motor-operated valves, A and B, were closed so that the flowmeter could be repaired, and afterwards the space between them contained air. Ethylene at a gauge pressure of 90 bar (1300 psig) was flowing along the 6 inch line into the 12 inch line. When the repair was complete valve B was opened, taking 3 seconds, thus allowing ethylene to flow into a length of pipe which contained only air at atmospheric pressure. The ethylene acted like a piston, compressing the air and

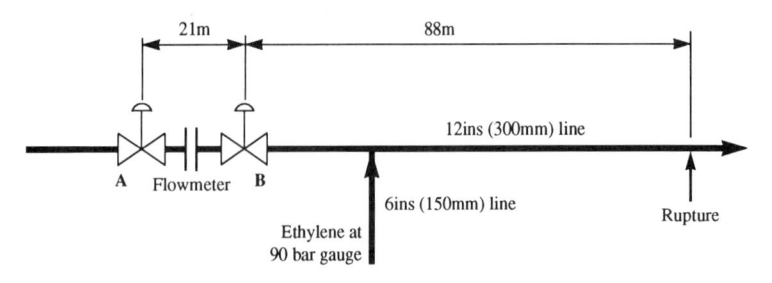

Figure 4.1 When valve B was opened, the ethylene acted like a piston and compressed the air between the two valves.

raising its temperature to 800°C. At this temperature the ethylene near the hot air started to decompose and the decomposition travelled along the pipeline. After the decomposition zone had travelled 88 m (290 feet) its speed became the same as that of the ethylene flow and all the decomposition took place in the same part of the pipeline. This heated the pipeline and it burst[9].

The amount of air in the pipeline was not sufficient for an air-ethylene explosion. The pipe would still have overheated and burst if it had contained nitrogen instead of air.

The compression of one gas by another was well known at the time[10,11], but not to pipeline operators. Once they know the hazard, they can easily prevent another incident by bringing pipelines up to pressure slowly.

4.1.5 COMMON MODE ISOMORPHISM

Toft[12] uses this phrase to describe situations in which different organisations, not normally in touch with each other, use the same materials. As an example he quotes the use of flammable foams in domestic furniture, railway sleeping cars (see Section 4.5.3, page 57), aircraft and hospitals and gives examples of fires that have occurred in all these locations. The best-known is the Manchester airport fire in 1985 where, 'the plane landed safely after take-off and sixty people died in the tail of the plane ... What killed them was not petrol — that started the fire. What killed them was the upholstery in the aeroplane'. After the fire aircraft seats were upgraded to better resist fire but this could have been done before the fire. The fact that foams produced toxic gases on exposure to fire was well known in other industries.

There is another example in Section 2.3 on page 17: vents can become choked in chemical plants and in the sewage systems of ships.

4.2 OTHER EXAMPLES OF FORGOTTEN KNOWLEDGE

4.2.1 SODIUM CHLORATE CAN EXPLODE

In 1977 there was an explosion and fire in a warehouse in which sodium chlorate was stored. According to the official report[13]:

'... the generally held view at that time, that commercially pure sodium chlorate would not be expected to detonate even under intense heat and conditions of confinement such as existed in this case, did not accord with the evidence of the explosion. The HSE therefore decided to undertake a research programme to investigate the behaviour of drummed sodium chlorate under fire conditions and reassess the explosibility of this material.

This decision was reinforced when enquiries revealed a previous similar explosion had occurred in Hamilton, Lanarkshire, in March 1969 ... Similar events involving either potassium or sodium chlorate were found to have occurred in St Helens in Lancashire in 1899, in Manchester in 1908, in Liverpool in 1910, in a Thames barge in Poplar in 1947 and in a cargo ship in Barcelona in 1974.'

Re-search is perhaps so-called because it so often re-discovers forgotten knowledge. The *Shorter Oxford Dictionary* defines research as, 'The act of searching (closely or carefully) for or after a specified thing or person. An investigation directed to the discovery of some fact by careful study of a subject; a course of critical or scientific inquiry'. There is no suggestion that the thing or fact must be discovered for the first time.

4.2.2 FIRE WATER CAN DISTURB POWDERS

In 1930 the Chemical Industries Association started a *Quarterly Safety Summary*. The first issue[14] included this advice:

'Attention is drawn to dust hazards which arise during fire-fighting ... from streams of water throwing up dust clouds when striking piles of powder.'

Fifty years later the following appeared in a report of a fire which damaged or destroyed 3000 tonnes of terephthalic acid powder[15]:

'Due to ... the disturbance of the powder by jets, there was throughout the incident an aerial ignition, giving at the height of the blaze a fireball 70 m high.'

4.2.3 SUDDEN MIXING CAN CAUSE A RUNAWAY

The following has often happened. A liquid is being added slowly to another already in a vessel. Someone notices that the stirrer or circulation pump is not running. He switches it on, the two liquids are mixed rapidly, react vigorously and are blown out of the vessel — or the vessel blows up. The two liquids do not have to react chemically; such incidents can occur when strong acids or alkalis are put into a vessel containing a water layer or when a cold volatile liquid is put on top of a hot one[16]. The following incident is typical of many. An aromatic hydrocarbon, 5-*tert*-butyl-*m*-xylene, was being nitrated in a 2 m^3 vessel, located on the first floor of a building, by slow addition of a mixture of nitric and sulphuric acids. After several hours the operator noticed that the temperature had not risen. He then noticed that the agitator had stopped. He switched it on. Almost at once he remembered that this could start a violent reaction. He switched off the agitator but it was too late. Within 20 seconds nitrous fumes were coming out of the vent on the vessel's lid. He went down to the ground floor to open the vessel's drain valve but the fumes were now too thick for him to see what he was doing. Wisely he decided to leave the area. Five minutes after he had switched on the agitator the vessel exploded and the contents ignited. The lid was found 15 m away and the rest of the vessel was propelled onto the ground floor[17].

It is possible to prevent such incidents by installing a trip so that agitator failure isolates the supply of incoming liquid. Note that it is not sufficient to measure the voltage or current supplied to the agitator motor. The agitation itself should be measured (or the liquid flow if there is an external circulation pump). For some processes, including nitration, a continous process which cannot run away, or is less likely to do so, is possible[18].

4.2.4 ENTRY TO CONFINED SPACES

Every year I read reports of people killed or injured while working inside vessels or other confined spaces — often in the chemical industry, often elsewhere[19]. Sometimes the confined space was not isolated from sources of danger, sometimes the atmosphere was not tested, sometimes training was poor. Some of the incidents occurred in backstreet companies, some in large, well-known ones. Here are two examples, picked at random from my files.

The first incident occurred in a large, well-known cable company. The contractors were laying new telephone cables in existing ducts, under contract to the telephone company. A 19-year-old youth was overcome by fumes and drowned at the bottom of a manhole, 7.5 m deep. The Health and Safety Executive found that the arrangements for monitoring for the presence of harmful gases were inadequate and that the standard of training and supervision was unacceptable. Both the contractor and telephone company were fined and the contractors were not allowed to carry out underground work anywhere until their training had improved[20].

Two of the dead man's fellow workers tried to rescue him and were themselves affected by the fumes; one was pulled out unconscious. This is a common feature of entry accidents. Even if they realise the risks, and often they do not, people naturally try to rescue others in distress. Whenever people have to enter confined spaces a stand-by man trained in rescue should be present and the necessary equipment should be alongside.

The second incident occurred on a ship. After carrying a load of 98% formic acid, the tanks were washed out with hot sea water for an hour and cold fresh water for half an hour. A member of the crew then entered one of the tanks to remove the remaining water. He was affected by gas. Another man entered the tank to rescue him but was himself affected, even though he was wearing breathing apparatus. Both men died.

Unknown to the crew, formic acid decomposes slowly to form water and carbon monoxide. This is, however, well known to manufacturers of formic acid who fit relief devices to drums containing the acid. No tests were carried out before the tank was entered. The second man was affected as he had a beard and this prevented a good seal between his face and the mask. The valve on the mask was of the demand type and the pressure inside the mask was low; men with beards or sideburns should wear masks fitted with valves which give an excess flow, not the minimum flow delivered by a demand valve[21].

4.2.5 HEAVY SUSPENDED LOADS CAN FALL

Many accidents have occurred because people were allowed to go underneath heavy suspended or temporarily supported objects. As far back as 1891 the UK House of Lords ruled that it was unsafe to allow a crane to swing heavy stones over people below[22].

In 1991 fourteen people were killed and nine seriously injured when a steel girder, 63 m long, weighing 53 tonnes, fell 10 m onto eleven cars which were waiting at traffic lights in Horoshima, Japan (Figure 4.2). The girder was being erected as part of a new railway line and was being raised on eight jacks, but the weight was not evenly balanced. According to the report, a crane would have been more suitable[23].

Figure 4.2 Ten cars were crushed and fourteen people killed when this girder collapsed. People should not be allowed beneath heavy suspended or temporarily supported loads.

4.2.6 LOW VOLTAGE TRANSFORMERS CAN GIVE A HIGH VOLTAGE SHOCK

In 1944 when I started work in industry I learned that some transformers, including the variable voltage ones widely used in laboratories, can be wired up wrongly. If an output of, say, 40 volts is wanted, instead of the terminals being at 0 and 40 volts they can be at 200 and 240 volts.

Recently a university professor connected up a transformer in this way. He touched the terminals to check that they were live (not dangerous at 40 volts) and got a severe shock.

4.3 LOSS OF SKILL

Skills can disappear with the passage of time, as the following incidents show.

4.3.1 LOSS OF MAINTENANCE SKILLS

MAINTENANCE OF SPECIALISED EQUIPMENT
A fire started with a leak of hydrogen and olefins from the body of a high pressure valve of an unusual design, operating at a pressure of 250 bar (3600 psi). The leak occurred at a spigot and recess joint between the two halves of the valve body; the gap between them was too large and some of the joint material was blown through it. The diametral clearance should have been 0.05 mm (0.002 inch) or less but in fact varied from 0.53 to 0.74 mm (0.021 to 0.029 inch) and was too large to contain the 0.8 mm ($1/32$ inch) compressed asbestos fibre gasket.

The valve concerned, and others like it, had been overhauled and repaired by a specialist workshop which had undertaken such work for decades. There were no written procedures but the 'old hands' knew the standard required. However, most of these men, including the supervisors and inspectors, had retired and standards had slipped.

We should never ask a workshop to simply 'overhaul', 'repair' or 'recondition' equipment but should agree with them precisely what is to be done and the standard required. Quoting the original Drawing or Standard is not sufficient if they do not specify allowable weardown tolerances.

Could we do more to debrief retirees? Professional employees might be encouraged to write their memoirs. Could other employees be encouraged to record theirs on tape, producing a sort of industrial oral history?

MAINTENANCE OF FLAMEPROOF ELECTRICAL EQUIPMENT
The source of ignition of the explosion described in Section 2.2, Incident 2 (page 11), may have been faulty flameproof electrical equipment as afterwards much

of the equipment on the plant was found to be faulty; screws were missing or loose, gaps were too large, plugs were missing and so on. Surveys elsewhere in the company showed a similar state of affairs. In one compressor house there were 121 items of which 33 were in need of some attention.

Further investigation showed that new electricians joining the company were not given any systematic training on the reasons why flameproof equipment is used, the standard of workmanship required and what can occur if that standard is not achieved. New men were merely asked if they knew how to maintain the equipment and if they said 'No' a supervisor gave them a quick demonstration. The standard of workmanship thus gradually fell.

There were other things wrong as well. Spare parts and the special tools required were often not in stock and there was no regular auditing of the maintenance standard. Flameproof equipment was sometimes used when cheaper, easier-to-maintain Zone 2 equipment would have been adequate.

The leak which ignited came from a badly-made high-pressure joint. Originally these joints were made only by an elite group of fitters who were trained to the high standard required. This was resented by the other fitters and so all were trained to the required standard, or so it was thought, and joints were made by whoever was available. Manpower utilisation was improved but not all the newly-trained men had the necessary skill, or perhaps they did not understand the need to use their skill to the full. After the explosion the original system was restored.

STANDARDS FORGOTTEN

In Section 2.2, Incident 1 (page 10), I describe a fire which started when the studs supporting the gland of a propylene injector, a high pressure (250 bar) reciprocating pump, failed by fatigue. Afterwards two safety studbolts were fitted to the injector, and others like it, to hold the gland follower in position if any of the four gland follower studbolts failed (Figure 4.3). The nuts on the safety studbolts were set 0.8 mm ($\frac{1}{32}$ inch) clear of the gland followers so that they were not subjected to pulsating loads.

Nearly 25 years after the original failure two gland follower studbolts failed. One of the safety studbolts also failed and there was a leak of hydrocarbon. This time, fortunately, it did not ignite.

The gland follower studbolts failed because they had been overtightened. The safety studbolt failed as the safety gap between its nut and the gland follower was too small or non-existent. Both surfaces were rough and the gap between them was no longer checked during maintenance.

SOME OTHER FAILURES TO LEARN FROM THE PAST

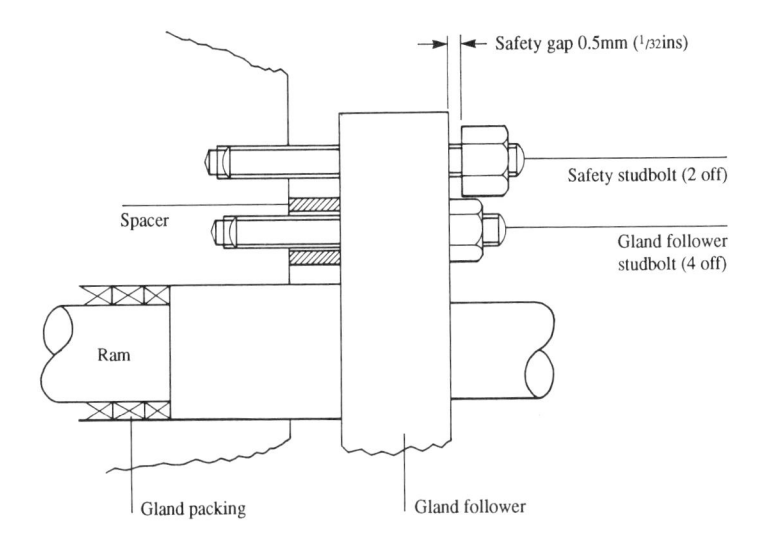

Figure 4.3   In this propylene injector gland, the safety gap was too small. The safety studbolt was subjected to pulsating loads and when the gland follower studbolts failed the safety studbolt also failed.

4.3.2   LOSS OF OPERATING SKILLS

- A runaway reaction occurred on a continuous plant; the design temperature of the reactor was 200°C but it reached 600°C. The report on the incident included the following:

'The reaction was believed by the present management, supervision and operating group to be readily controlled by simple procedures. The technical information to contradict this was well known to the licensor but, if known originally to us, had been diluted in handover between managers.

'Whilst the management, supervision and operators may once have been uniformly aware of the hazards of a temperature runaway, it is clear that successive changes of staff, coupled with familarity, have led to this reactor being regarded as a docile unit, occasionally 'lively' in the initial stages but always controllable. It is essential that the process be better documented to permit a more reliable technical handover between managers and that the methods by which the supervisors are made and kept familiar with the process be reviewed ... Reasons for instructions are not always given, critical processes are not highlighted from the routine ...'

American readers should note that the terms 'manager' and 'supervisor' do not have the same meaning in the USA as in the UK (see Preface).

- An explosion occurred in the drain line from a reactor, shattering the drain valve. During the reaction a solution of an unstable intermediate had leaked into the drain line and had then evaporated, leaving an unstable residue. When the reactor was heated, later in the batch, the heat may have been sufficient to detonate the residue.

  During the investigation someone noticed a small panel on the side of the drain valve. No-one knew what it was for. The manufacturer explained that it replaced a connection to a water line, originally used to wash out the drain line to remove any residue left after draining or leakage. Thirty years before the explosion there was a change in the processes used, the water line was no longer needed and it was removed. When unstable intermediates were again used all the people who had used the water wash had retired.

- The crash of a DC–10 aircraft in Chicago in 1979, in which 279 people were killed, was traced to deficiencies in quality control. An engine pylon flange was overstressed and failed during take-off, the engine falling to the ground. Afterwards similarly damaged flanges were found on another nine DC–10s. The deficiencies in quality control were due in part to the disruption caused when the assembly line was moved from one factory to another[24].

- Some oil-soaked pebbles between plant equipment were dug out and replaced by concrete. A scaffold pole standing on the pebbles was concreted in position and left there when the scaffolding was dismantled. A few years later, after the staff had changed, someone asked what the pole was for and was told that it was a vent on a drain. It was then fitted with a flame trap!

4.3.3 AN EXAMPLE FROM THE NUCLEAR INDUSTRY

In 1969 an experimental nuclear reactor at Lucens, Switzerland, overheated and was seriously damaged, though the escape of radioactivity was negligible. The reactor was cooled by carbon dioxide gas and the incident started when some of the water supplied to a compressor seal leaked into the gas stream. This caused corrosion and the corrosion products obstructed some of the channels in the fuel assemblies. The flow of carbon dioxide through these channels fell, the magnesium fuel cladding started to melt, causing further blockages and further overheating.

There were a number of weaknesses in the design but the initiating event was the the seal failure. The seals were well designed (by the standards of the 1960s) and the original supply had been thoroughly tested. Unfortunately the workman who had produced the required surface finish had moved on, his firm had been absorbed into a bigger company and by 1968 the original supply of seals had been used up[25].

4.4 TWO FIRES AND AN EXPLOSION IN AN OIL COMPANY
An official report[26] described two fires and an explosion that occurred between March and June 1987 in a major oil company. The report is thorough but does not point out that similar incidents had occurred elsewhere and that most of the recommendations had been made before.

INCIDENT 1
In the first incident, in which two men were killed, liquid accumulated in a dip in a blowdown line and leaked out when the line was opened (see Figure 4.4).

- A 1963 paper (reprinted in Reference 27) described a major fire, the result of liquid collecting in a dip in a blowdown line. The plant involved in the recent incident contained a small-bore line to drain liquid from blowdown lines but it was blocked by scale. A warning that small-bore lines can block in this way was included in the company's piping code (Paragraph 32 of the official report).

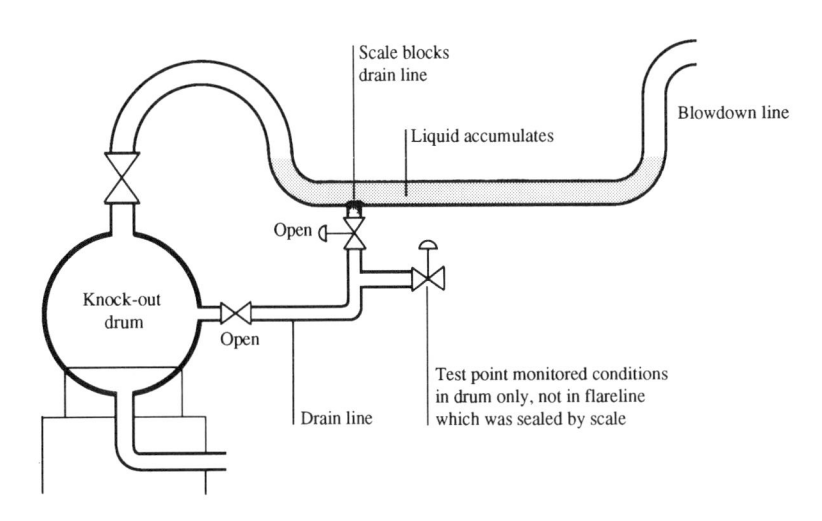

Figure 4.4 A drain line became blocked and liquid accumulated in a dip in a blowdown line. (Reproduced by permission of the Health and Safety Executive.)

- The liquid was present because isolation for maintenance depended on a large valve which leaked, and not on slip-plates. So many similar incidents have occurred that in 1980 I used them illustrate the theme that 'Organisations have no Memory'. Chapter 2 is based on that paper (see Preface).
- Access for the maintenance workers was poor. They 'had to crawl under or over the valve body for work on the other side' (Paragraph 19) and when a leak ignited they were 'engulfed in the fire' (Paragraph 26). A similar incident, though involving toxic gas, is illustrated in one of the Institution of Chemical Engineers' *Hazard Workshop Modules*[28], collections of slides and notes illustrating accidents that have occurred.

  Work on blowdown lines is particularly liable to result in spillages and in another incident a fitter escaped by sliding down a pipe like a fireman down a pole[29].
- The leak was ignited by a diesel engine. The fact that diesels can ignite leaks was widely publicised following a fire in 1969[30,31] (see Section 6.6, page 86). In the 1987 incident the exhaust gas spark arrestor was missing (Paragraph 35).
- The isolations necessary for maintenance were left to two supervisors to sort out shortly beforehand. Unlike the operational changes necessary they were not planned well in advance (Paragraphs 15–16). Reference 31 describes a similar incident that occurred many years ago.
- The joints were not broken in the correct manner: all bolts were removed and a slip-ring pulled out. Some bolts should have been left in and the joint opened gradually with a flange spreader (Paragraph 36). Again, a familiar story[28,32].

INCIDENT 2

The second incident, the explosion, was a spectacular one; a vessel burst and a piece, weighing 3 tonnes, was found 1 km away. Other pieces were thrown hundreds of metres (Figure 4.5). One man was killed. The cause was a classical one: liquid was let down from a high pressure vessel to a low pressure one. When the liquid level was lost, high pressure gas passed into the low pressure vessel; the relief valve was not big enough to protect the vessel and it burst.

There was a trip system to prevent overpressuring of the low pressure vessel but it was not in working order. Furthermore, as there were two let-down valves in parallel, a glance at the drawings without detailed calculations should have shown anyone familiar with reliability calculations that the hazard rate would be too high, even if the equipment were kept in working order. Reference 33 examines a similar situation in detail.

SOME OTHER FAILURES TO LEARN FROM THE PAST

Figure 4.5　This fragment of a low pressure vessel was blown 340 m. (Reproduced by permission of the Health and Safety Executive.)

INCIDENT 3
This was a fire in a crude oil tank while sludge was being removed. Vapour given off by the sludge was ignited by illegal smoking by contract cleaners who had removed their breathing apparatus so that they could smoke, and so that they could see better. Although the measured gas concentration was only 25% of the lower explosive limit, it is well known that the concentration can vary, especially when sludge is being disturbed[34]. The tank was not force-ventilated and no thought had been given to means of escape. One man, a non-smoker, was killed, perhaps because his airline became entangled with support pillars and hindered his escape. The incident shows once again the need for close supervision of contractors[35].

The purpose of this account is not, of course, to throw stones at the company involved, whose standards are normally very high, but to encourage constructive thought on ways of learning the lessons of the past. This was discussed briefly at the end of Chapter 2 and is discussed further in Chapter 10.

55

LESSONS FROM DISASTER

4.5 SOME INCIDENTS ON THE RAILWAYS

4.5.1   KING'S CROSS

The fire at King's Cross underground station, London, in 1987 killed 31 people and injured many more. The immediate cause was a lighted match, dropped by a passenger on an escalator, which set fire to an accumulation of grease and dust on the escalator running track. A metal cleat which should have prevented matches falling through the space between the treads and the skirting board was missing and the running tracks were not cleaned regularly.

No water was applied to the fire. It spread to the wooden treads, skirting boards and balustrades and after 20 minutes a sudden eruption of flame occurred into the ticket hall above the escalator. The water spray system installed under the escalator was not actuated automatically and the acting inspector on duty walked past the unlabelled water valves. London Underground employees, promoted largely on the basis of seniority, had little or no training in emergency procedures and their reactions were haphazard and uncoordinated.

Although the combination of a match, grease and dust was the immediate cause of the fire, an underlying cause was the view — accepted by all concerned including the highest levels of management — that occasional fires on escalators and other equipment were inevitable and could be extinguished before they caused serious damage or injury. From 1958 to 1987 there was an average of 20 fires per year, called 'smoulderings' to make them seem less serious. Some had caused damage, and passsengers had suffered from smoke inhalation, but no-one had been killed. The view thus grew that no 'smouldering' could become serious and fires were treated almost casually. (Compare the belief in Section 2.2, Incident 2 (page 11), that leaks would not ignite.) Recommendations made after previous fires were not followed up. Yet escalator fires could have been prevented, or reduced in number and size, by replacing wooden escalators by metal ones, by regular cleaning, by using non-flammable grease, by replacing missing cleats, by installing smoke detectors which automatically switched on the water spray, by better training in fire-fighting and by calling the Fire Brigade whenever a fire was detected, not just when it seemed to be getting out of control[36].

4.5.2   CLAPHAM JUNCTION

The immediate cause of this 1989 accident at Clapham in South London, in which 35 people were killed and nearly 500 injured, was wiring errors by a signalling technician. They were not detected as the checking and supervision were inadequate. One of the underlying causes was a failure to learn the lessons

SOME OTHER FAILURES TO LEARN FROM THE PAST

of the past. Similar wiring errors had been made a few years before, though the results were less serious.

New signalling had been installed on another line in 1986. There were three incidents in the following days. To quote the official report[37]:

'... two wires had been wrongly connected to the same terminal of a relay when there should have been only one wire. The similarities between the essential ingredients of this incident and what was to happen three years later at Clapham Junction need no further underlining ...

'They should have been used as clear pointers to what was going wrong in the installation and testing practices on Southern Region ...

'Instead, all these incidents produced was:
(i) A brief flurry of paperwork over three or four months which produced important information, but on a very limited circulation.
(ii) Despite an immediate recognition that the need for a new testing system was urgent, the emergence as late as eighteen months later of a document SL–53 which was never to be properly understood or implemented and was not to change any of the existing working practices for the better;
(iii) Despite the recognition of the importance of training staff, not a single training course on testing in the three years before the Clapham Junction accident; and
(iv) The appointment of the Regional Testing Team ..., with the objective of raising standards and preventing failures, but with a workload, a lack of resources, and a lack of management direction which meant that ... those objectives were never going to be met.'

The report summed up the response to the earlier incident with the words, '... the lesson was never properly learnt because it was never properly taught'.

Do not assume that the Clapham Junction accident occurred because standards of management were far worse than in other companies. Are you sure that similar failures never occur in your company?

4.5.3 TAUNTON: A FIRE IN A SLEEPING CAR

Twelve people were killed when a railway sleeping car caught fire in Taunton, Somerset in 1978. The car had been built in 1960 and fitted with steam heating. In 1978 this was replaced by electric heating and electric convector heaters were installed in the vestibules at the ends of the car. Bags of dirty linen from the

57

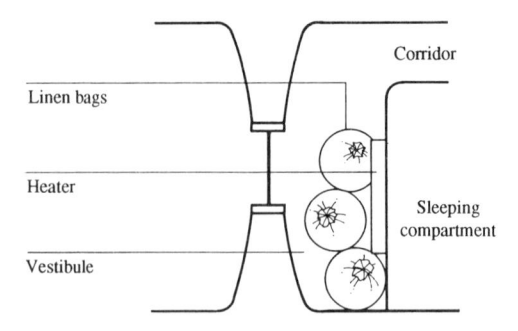

Figure 4.6 Bags of dirty linen, placed near the heater in the vestibule of a railway sleeping car, got too hot and caught fire.

previous journey were placed next to a heater as shown in Figure 4.6. The linen got too hot, caught fire and the fire spread down the coach.

A similar fire occurred on another line in 1973 but it was detected promptly and soon extinguished. Damage was slight, no-one was injured and no great attention was paid to the incident[38].

4.6   SOME SUCCESSES

To end this chapter, here are some accounts of occasions when someone remembered what had happened before. There must, of course, be many more such stories. Safety is the study of failures, not an account of the accidents that did not occur. We learn more from failures than from successes, this is why this book describes failures, but I would not like readers outside the process industries to think that we never learn. Many people do.

• During the night a shift fitter was asked to change a valve on a unit that handled a mixture of acids. He could not find a suitable valve in the workshop but after looking round he found one on another unit. He tested it with a magnet. It was non-magnetic so he assumed it was one of the valves made from grade 321 or 316 steel which were normally used and installed it. Four days later it had corroded and there was a leak of acid. The valve was actually made from Hastelloy, an alloy suitable for use on the unit where it was found but not suitable for use on the unit on which it was installed.

Three years later the same mistake was made again. This time a supervisor spotted the error in time. He not only remembered the original incident but had a copy of the report in his locker and also remembered the markings on the Hastelloy valve.

- While a man was dressing an abrasive wheel the tool he was using came apart and bits flew in all directions. Fortunately no-one was hurt. The design was poor and stocks of the tool were withdrawn from the stores.

  Six years later the same man was supplied with the old type of tool. A well-meaning person, who did not know its history, had put it back into stock.

What would have happened if the two key men involved in these incidents had left? We should not rely on people remembering but set up systems which reduce the chance of a repeat performance.

As discussed in Section 10.3 (page 170), on some plants they keep a 'history book' or 'black book' in the control room. It contains accident reports relevant to the plant's operations, both local reports and those from other plants and companies. It is compulsory reading for newcomers and others dip into it from time to time to refresh their memories. Have you got one?

In 1990 I visited a factory where an explosion had occurred 14 years before, fortunately without causing injury. While we were being shown round one of the units I asked the young manager if this was the site of the explosion. He was probably in his first year at university when it occurred but he explained clearly what had happened and why. This is no more than he should be able to do, you may say. But on many plants, as this book shows, people have little knowledge of incidents that happened ten or more years ago. It is good to know that in one factory, at least, the lessons of the past are now remembered.

REFERENCES IN CHAPTER 4
1. Health and Safety Executive, 1985, *The Abbeystead Explosion* (HMSO, London, UK), 20.
2. *Daily Telegraph*, 14 March 1987 and 19 February 1988.
3. Fowler, R., 1986, in *Research and Innovation in the 1990s*, edited by Atkinson, B. (Institution of Chemical Engineers, Rugby, UK), 59.
4. Parker, R.J. (Chairman), 1975, *The Flixborough Disaster: Report of the Court of Inquiry* (HMSO, London).
5. Kletz, T.A., 1988, *Learning from Accidents in Industry* (Butterworths, London, UK), Chapter 8.

6. Lees, F.P., 1980, *Loss Prevention in the Process Industries*, Volume 2 (Butterworths, London, UK), Appendix 1.
7. *Report of the Tribunal Appointed to Inquire into the Disaster at Aberfan on October 21st, 1966*, 1967 (HMSO, London, UK).
8. Kletz, T.A., 1988, *Learning from Accidents in Industry* (Butterworths, London, UK), Chapter 13.
9. McKay, F.F. et al, November 1977, *Hydrocarbon Processing*, 56 (11): 487.
10. Pritchard, H.O., 1960, *Quarterly Reviews*, 46.
11. Steinberg, M. and Kaskan, W.E., 1955, *Fifth Symposium on Combustion* (Reinhold, New York, USA), 664.
12. Toft, B., 1992, *Disaster Prevention and Management*, 1 (3): 48.
13. Health and Safety Executive, 1979, *The Fire and Explosion at Braehead Container Depot, Renfrew, 4 January 1977* (HMSO, London, UK), 13.
14. *Quarterly Safety Summary*, January–March 1930, 1 (1).
15. *Fire*, June 1980, 23.
16. Kletz, T.A., 1988, *What Went Wrong? — Case Histories of Process Plant Disasters*, 2nd edition (Gulf, Houston, Texas, USA), Section 3.2.8.
17. Kotoyori, T., January 1991, *Journal of Loss Prevention in the Process Industries*, 4 (2): 120.
18. *The Chemical Engineer,* February 1979, No 341, 79.
19. Kletz, T.A., 1988, *What Went Wrong? — Case Histories of Process Plant Disasters*, 2nd edition (Gulf, Houston, Texas, USA), Chapter 11.
20. *Health and Safety at Work*, June 1989, 11 (6): 11.
21. *Loss Prevention Bulletin*, April 1984, No 056, 24.
22. Farmer, D., November 1986, *Health and Safety at Work*, 8 (11): 61.
23. Yasuda Fire and Marine Insurance Co, April 1992, *Safety Engineering News* (Japan), No 17, 7.
24. Stewart, S., 1986, *Air Disasters* (Ian Allen, London, UK), 160.
25. Mosey, D., 1990, *Reactor Accidents* (Butterworths, London, UK), 53.
26. Health and Safety Executive, 1989, *The Fires and Explosion at BP Oil (Grangemouth) Refinery Ltd* (HMSO, London, UK).
27. Laney, T.J., 1985, in *Fire Protection Manual for Hydrocarbon Processing Plants*, Volume 1, edited by Vervalin, C.H., 3rd edition (Gulf Publishing, Houston, Texas, USA), 101.
28. *Hazard Workshop Module No 004, Preparation for Maintenance* (Institution of Chemical Engineers, Rugby, UK), undated.
29. Kletz, T.A., 1988, *What Went Wrong? — Case Histories of Process Plant Disasters*, 2nd edition (Gulf Publishing Co, Houston, Texas, USA), Section 10.4.3.
30. *Chemical Age*, 12 December 1969, 40 and 9 January 1970, 11. Similar reports appeared in many other journals.
31. Kletz, T.A., 1988, *Learning from Accidents in Industry* (Butterworths, London, UK), Chapter 5.
32. Kletz, T.A., 1988, *What Went Wrong? — Case Histories of Process Plant Disasters*, 2nd edition (Gulf Publishing Co, Houston, Texas, USA), Section 1.5.1.

33. Kletz, T.A. and Lawley, H.G., 1982, in *High Risk Safety Technology*, edited by Green, A.E. (Wiley, Chichester, UK), 317.
34. Fawcett, H.H. and Wood, W.S. (editors), 1982, *Safety and Accident Prevention in Chemical Operations* (Wiley, New York, USA), 820.
35. Kletz, T.A., 1988, *Learning from Accidents in Industry* (Butterworths, London, UK), Chapter 16.
36. Fennell, D. (Chairman), 1988, *Investigation into the King's Cross Underground Fire* (HMSO, London, UK).
37. Hidden, A. (Chairman), 1989, *Investigation into the Clapham Junction Railway Accident* (HMSO, London, UK), Sections 9.50–53 and 9.59.
38. *Report on the Fire that Occurred at Taunton in the Western Region of British Railways*, 1979 (HMSO, London, UK).

5. WHY ARE WE PUBLISHING FEWER ACCIDENT REPORTS?

'A person is held responsible for the sins of his family, of his community, or even of all mankind, when he fails to use his influence for the correction of wrongs.'
Mishnah (code of Jewish law), 2nd century AD

5.1 THE PRESENT POSITION

The first step in learning from the past is to make the information available, both in our own companies and elsewhere. In fact, the oil and chemical industries are now publishing fewer reports on accidents that have occurred than in the 1970s. This chapter discusses the reasons why we are not publishing, the reasons why we should publish and the action needed to improve publication.

Is it true that we are publishing less? The American Institute of Chemical Engineers (AIChE) held their first Loss Prevention Symposium in 1967 and they have been held at roughly annual intervals since then. If we compare the papers presented at the first five symposia (published in *Loss Prevention,* volumes 1–5) with those presented in recent years (published in *Plant/Operations Progress*), we see that the early symposia contained many accounts of accidents that had occurred and the actions recommended to prevent them happening again. Now such papers are few; most papers describe *equipment* for making plants safer, *equations* for calculating effects or the results of *experiments*. (There are also many papers on procedures for identifying and assessing hazards.) Important as these three Es are, they do not have the same impact on reader or audience as descriptions of accidents that have occurred. Nothing else, except an accident in which one is personally involved, is as effective as a stimulus to action.

The AIChE experience is mirrored elsewhere. The UK Chemical Industries Association and the US Chemical Manufacturers Association no longer publish case histories. The UK Institution of Chemical Engineers *Loss Prevention Bulletin* used to consist mainly of incident reports; now they make up only a part of the contents. The number of case histories presented at conferences on loss prevention is small. Instead the emphasis is on the three Es. When I was a safety adviser with ICI (1968–1982), I used to send hundreds of copies of a monthly *Safety Newsletter* containing accounts of accidents to many other companies throughout the world; this stopped soon after I retired. (My book *What Went Wrong?*[1] contains many extracts from the *Newsletters*.)

Before we discuss the reasons for these changes, it may be useful to discuss the reasons why we should spread information on accidents that have occurred.

5.2   WHY SHOULD WE PUBLISH ACCIDENT REPORTS?

(1) The first reason is moral. If we have information that might prevent an accident then we have a duty to pass on that information to those concerned. If there is a hole in the road and we see someone about to fall into it, we have a duty to warn him. If we fail to do so we are, to some extent, responsible for his injury. The 'holes' into which oil and chemical plants fall are more complex than a hole in the road but the principle is the same.

(2) The second reason is pragmatic. If we tell other people about our accidents, then in return they may tell us about theirs, and we shall be able to prevent them happening to us. However, we cannot expect a precise *quid pro quo*. Inevitably the bigger companies will give more than they receive.

If we use the information on accidents published by other companies, but do not make our own information available, we are 'information parasites'. Biologists use this phrase to describe those birds, for example, that rely primarily on neighbouring birds to give warnings that enemies are approaching[2].

(3) The third reason is economic. Many companies spend more on safety measures than some of their competitors and thus pay a sort of self-imposed tax. If we tell our competitors about the action we took after an accident, they may spend as much as we have done on preventing that accident happening again. Responsible companies often welcome new legislation which compels everyone to adopt safety measures that they cannot afford to adopt unilaterally. In 1862 the UK Parliament passed the Alkali Act which prohibited the discharge of hydrogen chloride to the atmosphere. According to Brimblecombe[3]:

'The bill did not prove a very controversial one, partly because the alkali manufacturers admitted that the industry was responsible for considerable environmental damage and were already tired of being the subject of numerous complaints ... Their reason for refusing to control the emissions from their plants was merely one of competitive effectiveness.'

(4) The fourth reason is that if one company has a serious accident, the whole industry suffers in loss of public esteem while new legislation may affect the

whole industry. As far as the public and politicians are concerned, we are one. Flixborough cost the insurance companies about £70 million in material damage. It cost the rest of the chemical industry much more. The same is true of Seveso and Bhopal. To misquote the well-known words of the English poet, John Donne:

'No plant is an Island, entire of itself; every plant is a piece of the Continent, a part of the main. Any plant's loss diminishes us, because we are involved in the Industry: and therefore never send to know for whom the Inquiry sitteth; it sitteth for thee.'

Note that the usefulness of case history reports is increased if:
(a) they give information, preferably quantitative, on the damage caused so that other people are able to predict damage; and
(b) they contain sufficient information for people with different interests, experience and background to draw additional conclusions from the incident[4] (see Section 8.13, page 140).

5.3   WHY ARE WE PUBLISHING FEWER ACCIDENT REPORTS?

(1) The first and most important reason is time. Since the recession of the early 1980s many companies have reduced staff; those who are left have less time for putting accident reports into a form suitable for publication, or attending conferences to present them. When people are under pressure, jobs that they intend to do and know they ought to do but do not have to do by a specific time, get repeatedly postponed. This is especially true of those jobs which bring them less credit with their employers than attending to output and efficiency.

(2) The second reason is the influence of company lawyers, particularly in the USA where they are much more powerful than in the UK (Figure 5.1). They often say that publication of an accident report or even circulation within the company may effect claims for compensation or lay the company open to prosecution. Huber writes, 'Tort litigation is definitely a supply-side industry. Its growth has been driven by the availability of information about hazards, not by the severity of the hazards themselves'[5].

In the UK the lawyer is more often an excuse for non-publication than a reason. Lawyers usually withdraw their objections to publication if we insist that otherwise someone else is going to get hurt and they will be quoted as the reason why nothing was done. It is usually possible to disguise an incident

WHY ARE WE PUBLISHING FEWER ACCIDENT REPORTS?

Figure 5.1 For a number of reasons companies, especially in the USA, are reluctant to broadcast information on the accidents they have had.

sufficiently to avoid legal problems. For example, when a hired crane collapsed on a plant there was a dispute about responsibility but the plant lawyer agreed that a photograph could be published if the name of the crane owner was blacked out (see Section 8.9, page 136).

In the USA, lawyers are more powerful and the legal system is different. Nevertheless some US engineers seem able to stand up to their company lawyers and insist that accident prevention should not always take second place to the protection of the company's profits.

(3) Another reason is fear of adverse publicity. Ingles writes[6], '... one's superior authorities — especially if a large public authority — are extremely sensitive, to the point of phobia, about the publication of anything that went wrong'. If an

65

incident is described at a conference or in a technical journal, perhaps the press or TV will pick it up and use the incident to show that the company is careless, irresponsible and so on. This does not seem to happen in practice. If it did, the company could say that publication showed their sense of responsibility.

However, if fear of adverse publicity is not a valid reason at the company level, it is powerful at the individual or group level. People are naturally reluctant to draw attention to their failures and expose their weaknesses, particularly if the company culture is intolerant of error. I have heard someone say, 'A man who reports an accident is a double fool: once for having it and once for reporting it'. In these circumstances an unholy alliance of engineers and lawyers can agree to suppress an accident report.

(4) In many companies the procedures to be followed in order to publish are so laborious that they discourage publication. Many departments have to be consulted; each one takes its time and may justify its involvement by suggesting changes. Time has to be spent following a paper around to make sure it has not reached the Pre-Cambrian layer at the bottom of someone's intray. In contrast, to show what is possible in some companies, during my 14 years as a safety adviser with ICI I published about 70 papers. Each one had to be approved only by the Patents and Agreements Department, the Publicity Department and a Division Director. It was then reported to the Division Board of Directors, but that was a formality. The whole approval procedure rarely took more than a month, often less. Of course, I often discussed my papers with colleagues in other departments but they were not involved in the official approval procedure.

(5) It is sometimes said that publication of accident reports may result in disclosure of confidential information. These fears are unfounded. Inessential details can be altered to avoid giving away secrets without destroying the essential safety message. (However, to preserve confidentiality a colleague of mine halved all the pressures when writing a paper. For the same reason the Patents and Agreements Department then doubled all the pressures!)

(6) A final reason why fewer reports are being published is the retirement of several people who were personally committed to the spread of information, who published a great deal and who exchanged other reports unofficially. Of course, they could do so only because they had the support of their companies, but the commitment in their companies was not strong enough to survive their departure.

5.4 HOW CAN WE ENCOURAGE THE TRANSFER OF INFORMATION?

(1) Encouragement from the top is of primary importance. Statements of policy count for little. A note from a senior manager saying, 'I liked your paper on the fire we had last year' or 'Are you going to publish the report on the gas leak?' is much more effective. A few committed individuals may publish despite difficulties and lack of encouragement; most people will not do so unless encouraged by their bosses.

(2) Companies, and particularly their lawyers, are usually more willing to allow publication of reports on near-misses, in which no-one was injured, than reports on accidents that resulted in injury or death. We can learn just as much from these near-misses as from other accidents.

(3) Anonymity helps. Accidents do not reflect credit on those concerned and no-one likes to see their mistakes publicised. Accident reports — even for internal company circulation — should, whenever possible, be edited so that the plant or factory where the accident occurred cannot be identified. This is easy in large companies but difficult in small ones.

If you do not want a report to be published under your own name, send it to a journal that will publish it anonymously, for example, the *Loss Prevention Bulletin* (published by the UK Institution of Chemical Engineers, 165–171 Railway Terrace, Rugby, CV21 3HQ, UK).

There are a number of clubs where accident reports are discussed in confidence. They are open to companies manufacturing particular chemicals such as chlorine, vinyl chloride or polyethylene. A club originally restricted to companies engaged in hydrocarbon oxidation has now widened its activities and become the International Process Safety Group. Clubs like this one provide a valuable mechanism for the circulation of accident reports to many companies which would not otherwise receive them but in an ideal world they would not be necessary; everything worth publishing would be in the open literature.

(4) People will be more willing to publish accident reports if we can establish a blame-free atmosphere. We should do so for several reasons. First, if people think they will be blamed for the errors that caused an accident, they will not tell us everything that happened and we shall not be able to prevent it happening again. An indulgent attitude towards people who have a momentary lapse of

attention, make an error of judgement or do not always follow the rules is a price worth paying for information on the cause of the accident.

Second, as shown in Section 7.7 (page 106), most human errors arise out of the work situation, not out of a deliberate decision. We cannot prevent them by telling people to be more careful, only by changing the plant design or method of working, improving the training or instructions or providing better supervision.

Third, responsibility is widely spread among many people. To quote from the *Robens Report*, the report which led to the Health and Safety at Work Act 1974[7]:

'The fact is — and we believe this to be widely recognised — the traditional concepts of the criminal law are not readily applicable to the majority of infringements which arise under this type of legislation. Relatively few offences are clear cut, few arise from reckless indifference to the possibility of causing injury, few can be laid without qualification at the door of a single individual. The typical infringement or combination of infringements arises rather through carelessness, oversight, lack of knowledge or means, inadequate supervision or sheer inefficiency. In such circumstances the process of prosecution and punishment by the criminal courts is largely an irrelevancy. The real need is for a constructive means of ensuring that practical improvements are made and preventative measures adopted.

(5) If you cannot publish the report on an accident, perhaps you can publish details of the action you took as a result. This may not have the same impact as the report, but it is a lot better than nothing. For example, in 1977 reports appeared in the technical press[8,9] of a fire on a large storage tank containing refrigerated propane. The company concerned said that for legal reasons they could not publish a report on the incident and, so far as I am aware, no report has ever appeared. But an employee of the company did present several conference papers describing new standards for cryogenic storage[10–12]. It is not difficult to read between the lines and see what probably happened[13].

(6) Finally, stand up to the company lawyer. Ask him if he really wants to let people fall into that hole in the road. It is easy for an Englishman to say this, for in the UK the company lawyer is a much less powerful figure than in the USA. Nevertheless, as I have already said, a few US engineers do stand up to their lawyers and their companies have not (at least, so far) become bankrupt. It is possible, however, that US companies will not be able to publish as widely as UK ones without a change in American tort law[5].

REFERENCES IN CHAPTER 5

1. Kletz, T.A., 1988, *What Went Wrong? — Case Histories of Process Plant Disasters*, 2nd edition (Gulf, Houston, Texas, USA).
2. Burger, J. and Gochfeld, M., June 1992, *Natural History*, 45.
3. Brimblecombe, P., 1988, *The Big Smoke* (Routledge, London, UK), 137.
4. Kletz, T.A., 1986, *Proceedings of the 5th International Symposium on Loss Prevention and Safety Promotion in the Process Industries* (Société de Chimie Industrielle, Paris, France), 20–1.
5. Huber, P.W., 1986, The Bhopalization of American tort law, in *Hazards: Technology and Fairness* (National Academy Press, Washington DC, USA), 89.
6. Ingles, O.G., July–September 1980, *Safety in Civil Engineering*, 1 (1): 15.
7. *Safety and Health at Work: Report of the Committee 1970–1972 (The Robens Report)*, 1972 (HMSO, London, UK), paragraph 261.
8. *Middle East Economic Digest*, 15 April 1977.
9. *Oil and Gas Journal*, 18 April 1977, 29.
10. Cupurus, N.J., 1979, Cryogenic storage facilities for LNG and NGL, *Proceedings of the 10th World Petroleum Congress* (Heyden, London, UK), 119 (Panel Discussion 17, Paper 3).
11. Cupurus, N.J., 1980, Developments in cryogenic storage tanks, *6th International Conference on Liquefied Natural Gas, Kyoto, Japan, 1980*.
12. Cupurus, N.J., 1981, Storage of LNG and NGL, *Seminar on LNG and NGL in Western Europe in the 1980s, Oslo, 2 April 1981*, Session II.
13. Kletz, T.A., 1988, *What Went Wrong? — Case Histories of Process Plant Disasters*, 2nd edition (Gulf, Houston, Texas, USA), Section 8.1.5.

6. WHAT ARE THE CAUSES OF CHANGE AND INNOVATION IN SAFETY?

> *'Men wiser and more learned than I have discerned in history a plot, a rhythm, a predetermined pattern. These harmonies are concealed from me. I can see only one emergency following upon another, as wave follows upon wave, only one great fact with respect to which, since it is unique, there can be no generalisations, only one safe rule for the historian: that he should recognise in the development of human destinies the play of the contingent and the unforeseen. This is not a doctrine of cynicism and despair. The fact of progress is written plain and large on the page of human history; but progress is not a law of nature. The ground gained by one generation may be lost by the next.'*
> H.A.L. Fisher, 1936, *A History of Europe*, Preface.

As we have seen in earlier chapters, serious accidents are a major cause of change in safety, even though the change is not always sustained. The greater the loss of life, the damage and the consequential loss, the greater the probability that change will result. However, serious accidents are not the only cause of change. In this chapter I discuss five significant innovations with which I have been involved. Two took place as a result of major incidents, one in the company and one elsewhere; two took place after a problem was recognised, without waiting for a serious accident to occur. The fifth was a change of a rather different nature, and followed a series of incidents rather than a specific event.

Change is a wider term than innovation. Most changes consist of the adoption of measures long recommended. New ideas are introduced only occasionally.

The changes described below took place in the Petrochemicals (formerly Heavy Organic Chemicals) Division of ICI mainly during the period 1967–1982. Similar changes took place, to varying extents, in other companies.

6.1 ISOLATION FOR MAINTENANCE

This is an example of a change brought about quickly by a single traumatic experience: Incident 2 in Section 2.1 (see page 6). Readers will recall that while a pump was being dismantled for repair, hot oil above its auto-ignition temperature came out and caught fire as the suction valve had been left open. Three men were killed and the unit was destroyed.

After the fire, instructions were issued by the production director, with the backing of the Division Board, that in all of the Divisions's works, not just in the one where the fire occurred, the following action must be taken before any equipment is handed over to the maintenance organisation:

(1) The equipment must be isolated by slip-plates (blinds) (or physical disconnection and blanking) unless the job to be done is so quick that fitting slip-plates (or disconnecting) would take as long as the main job and be as hazardous.

(2) Valves used to isolate equipment for maintenance, including isolation for slip-plating (or disconnection), must be locked shut with a padlock and chain or an equally effective device.

(3) If there is a change in intention — for example, if it is decided to dismantle a pump and not just work on the bearings — the permit-to-work must be handed back and a new one issued.

It was recognised that exceptions to these rules might be necessary but they had to be authorised in writing by the assistant works manager and reviewed annually.

It was most unusual for the Board to issue instructions of this sort. In the past such matters had been left to the four works managers, each of whom was responsible for about a thousand employees. While the Division was in a state of shock, following the fire, it was possible to impose a new rule by edict, with only the minimum of discussion.

There was little or no opposition to the new rules (really a revival of old rules but the earlier 1928 incident, described in Section 2.1 (page 4), had been forgotten). However, their implications were greater than had been foreseen. On many existing plants it was found difficult to insert slip-plates, as there was insufficient spring in the pipework. Sometimes only thin slip-plates could be used which were not capable of withstanding the same pressure as the pipework. The engineering department agreed to fit slip-rings or spectacle plates in new plants when there was insufficient spring for the insertion of slip-plates.

Some valves, such as ball valves, were difficult to lock off and brackets of various types had to be devised.

Fitting slip-plates is itself not entirely free from hazard and the degree of isolation necessary had to be agreed. Figure 6.1 summarises the code that was developed, after several years of discussion.

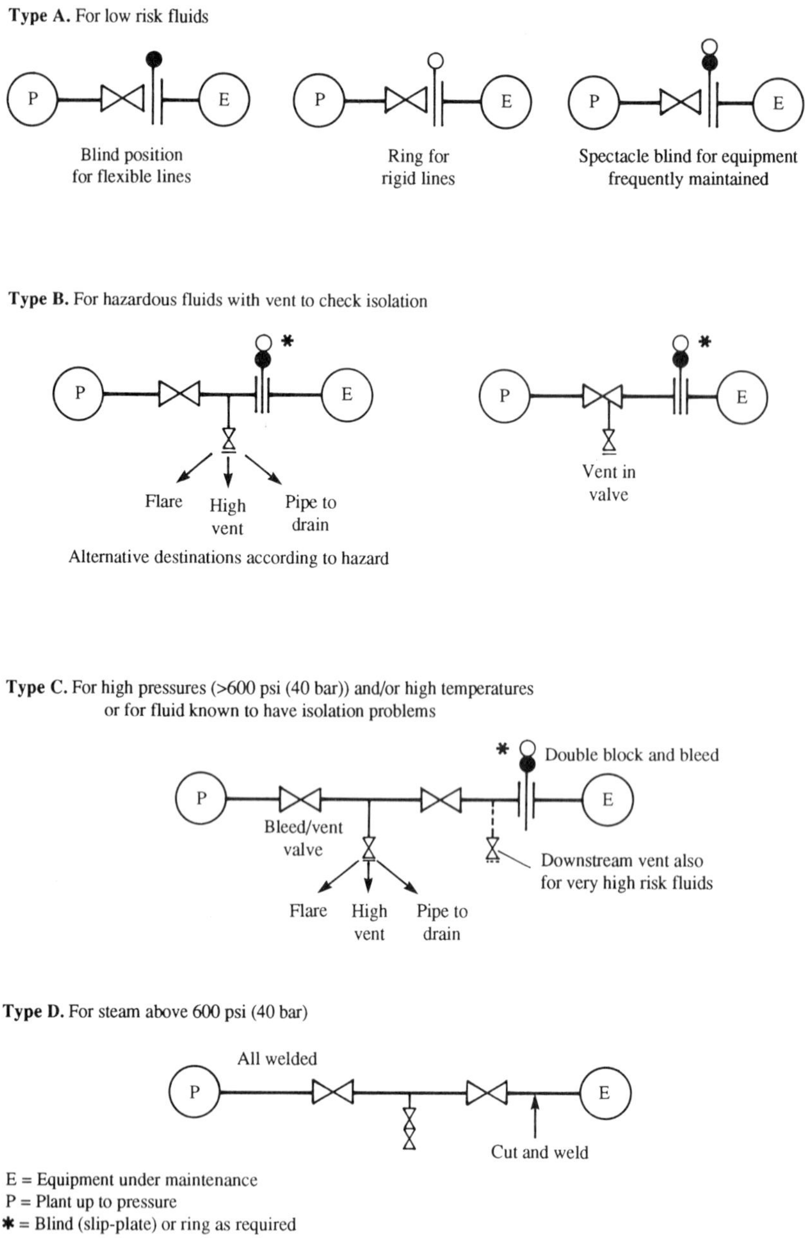

LESSONS FROM DISASTER

Type A. For low risk fluids

Blind position for flexible lines

Ring for rigid lines

Spectacle blind for equipment frequently maintained

Type B. For hazardous fluids with vent to check isolation

Flare   High vent   Pipe to drain

Alternative destinations according to hazard

Vent in valve

Type C. For high pressures (>600 psi (40 bar)) and/or high temperatures or for fluid known to have isolation problems

Double block and bleed

Bleed/vent valve

Flare   High vent   Pipe to drain

Downstream vent also for very high risk fluids

Type D. For steam above 600 psi (40 bar)

All welded

Cut and weld

E = Equipment under maintenance
P = Plant up to pressure
✱ = Blind (slip-plate) or ring as required

Figure 6.1   Methods of isolating equipment for maintenance.

By 1969 it was necessary to consider what should be done to make sure that the rules did not lapse (again). As we have seen, memories soon fade, not in the works where a major incident occurs, but elsewhere. The need for a continuing effort was emphasised two years later when a reorganisation doubled the size of the Division and brought into it many people who did not have vivid memories of the fire.

The ongoing activities took three forms:

- The operating and design staff of the Division met in groups of 12–20 to discuss the fire. They questioned the discussion leader (usually me) to establish the facts and then went on to say what they thought ought to be done to prevent similar incidents happening again. They worked out for themselves the need for the rules imposed after the fire. Each discussion lasted 1–2 hours and the rest of the morning was taken up by a discussion of other incidents involving preparation for maintenance. The discussions were repeated at intervals for new staff and about five years later many established staff attended again to refresh their memories. In some works, foremen and operators attended similar discussions.

Section 10.2 (page 167) describes these discussions in more detail.

- Plant managers (the equivalent of supervisors in the USA), more senior managers and safety advisers made regular checks of permits-to-work to see that they were completed correctly and that the precautions specified on the permits were actually carried out. The errors found were mainly due to misunderstandings rather than corner-cutting. In addition, safety surveyors from the Division headquarters visited each works two years after the fire, and again after another three or four years, to survey the way in which equipment was prepared for maintenance; detailed reports with recommendations were sent to each works manager and the production director.

- References to the fire and its cause and to similar incidents elsewhere (of which, regretfully, there was and still is no shortage; see Section 2.1, page 9) were included in the Division's monthly *Safety Newsletter*.

Everyone (at least, everyone of my generation) knows the stories of Adam and Eve, Noah's Ark, David and Goliath and many more, as we were repeatedly told them from an early age. I am not suggesting that loss prevention should be taught in infant schools but I do suggest that university chemical engineering students should be told about some of the accidents that have occurred and the action needed to prevent them. Further, we should be reminded of them from time to time throughout our careers.

A common experience after an accident is that a change is made in the plant or works where it occurred but not elsewhere. The rules made after the fire

LESSONS FROM DISASTER

applied to the whole of the Division. The Division Board had no authority to impose them elsewhere, but the other ICI Divisions were kept fully informed. The extent to which they changed their own practices was very variable. At the time ICI was very decentralised and the company Board left technical matters to the Divisions.

I hope that younger readers will not dismiss the fire, and the other incidents described in this chapter, as ancient history, of no relevance to the present day. If your company follows the procedures I have described, or similar ones, then you should know why these procedures were introduced. It may discourage you, and your colleagues, from scrapping them to save time. If your company does not follow the procedures, or similar ones, then you have been warned what to expect. I still read frequent reports of accidents that would not have occurred if the lessons of the fire had been learnt.

For further thoughts on the fire see Section 7.4, page 99.

6.2 TRIP AND ALARM TESTING

This is an example of a change resulting from recognition of a problem rather than from a serious incident or series of incidents.

Before the 1939–45, war chemical plants were mainly hand controlled. (The first plant on which I worked made *iso*-octane from butane; when constructed in 1940, it contained more automatic control than any other plant on the Billingham site.) After the war, during the 1950s and 1960s, the number of automatic trips and alarms grew greatly. The growth was technology led — that is, the trips and alarms were installed because they were now available — but they also satisfied demands for better efficiency, higher safety and fewer operators.

It was soon realised that trips and alarms sometimes fail to operate when required and minor incidents had occurred as a result. What factors affect the probability that a trip or alarm will fail?

The most important factor is regular testing. The figures shown in Table 6.1 were produced to show the effect of test interval on hazard rate, the rate at which dangerous incidents occur, for a typical trip (failure rate assumed: twice in three years). The *demand* rate is the rate at which the temperature, pressure, level or whatever parameter is being measured reaches the set point. The *hazard* rate is the rate at which it exceeds the set point without the trip operating. Two demand rates, once a year and once in five years, are considered.

The calculation is approximate as no account has been taken of common mode failures or time off-line for testing. In addition, the failure rate of trips is now somewhat lower than the figure assumed. Nevertheless, the figures are

TABLE 6.1
Effect of test frequency on hazard rate (assuming a failure rate of twice in three years).

Demand rate	Once a year	Once in 5 years
Hazard rate with		
weekly testing	Once in 150 years	Once in 750 years
monthly testing	Once in 36 years	Once in 180 years
annual testing	Once in 4 years	Once in 15 years

reproduced in their original form as they provided the incentive for the actions described below.

The main effort was devoted to surveys. Over the period 1969–1972 two surveyors witnessed the routine testing of every trip and alarm in the seven major works of ICI Petrochemicals Division. They looked to see that:

- the trip was tested at a reasonable interval, usually once/month but sometimes more often, sometimes once/quarter;
- the test was thorough and simulated 'real life' conditions;
- any faults found were corrected;
- the equipment was put back on-line after testing.

The results of the survey were discussed with local managers and detailed reports were prepared for the works managers and the production director. By the time these reports were issued it was possible to say that most of the recommendations were complete or in hand.

The findings of the survey varied widely. On some plants only a few minor faults were found in the procedures and these were quickly and willingly corrected. On most plants there was a handful of faults. On a few plants little testing was done and many trips and alarms were out-of-order.

Most of the faults found could not be put right overnight. It took several years, in some cases, to establish test routines and catch up with the backlog of maintenance. Extra staff had to be recruited and trained. Some people with long experience had to change their ideas. Many trips were removed; as they had never been kept in working order, the need for them was questioned.

Instrument staff on the whole recognised the need for testing and were keen to see that it was carried out. Some process staff were lukewarm; they seemed to feel that the desire of the Instrument Section to test 'their' instruments interfered with production. They were slow to realise that testing is necessary for the safety of the plant and slow to realise that they, not the Instrument Section, should own the trips. Slowly most people were persuaded to change their views. Everyone recognised the need to test relief valves and failure to do so was rare. Slowly people came to realise that failure of a trip can be just as serious as failure of a relief valve, and that testing of trips is just as important as relief valve testing. In some companies this is not recognised even today and trip testing has a lower profile than relief valve testing.

A repeat survey, in which only a proportion of the trips and alarms were looked at, was carried out during 1975 and 1976.

At no time was there a directive from the Board or anyone else outside the works that 'All trips must be tested regularly'. Progress was made by pointing out faults and their likely consequences and by persuading operating staff to make the necessary changes. (Safety advisers should be persistent but they also have to be good-natured or they will get their clients' backs up.) The Board was supportive; they allowed the necessary resources to be made available and this, of course, was known. After a few years a Division Code on testing was drawn up but this merely required people to do what, by this time, they were already doing. It was an example, at company level, of statute law defining what had already become 'custom and practice', the common law of the organisation. The procedure followed was thus quite different from that used in introducing a new permit-to-work system (Section 6.1, page 70) where, after an accident, new rules were imposed by decree without much persuasion or consultation.

Company cultures differ and in some organisations it might be appropriate to start with a directive or statement of policy from the Board. However, we live in a society in which people are reluctant to carry out instructions unless they are convinced that they are necessary. Whether or not the Board issues directives, we still have to convince people that change is needed, so why bother with the initial decree from on high?

After a serious accident people will accept directives but even then, when the shock has worn off, we may have to convince people that the new procedure is the best one.

6.3   STORAGE OF LIQUEFIED PETROLEUM GAS

This is an example of a change resulting mainly from a serious incident in another company. At Feyzin, France, in 1966 a leak of propane, a liquefied petroleum gas (LPG), caught fire. One of the LPG storage vessels was heated by the flames and burst, a phenomenon known as a BLEVE (Boiling Liquid Expanding Vapour Explosion). Fifteen to eighteen people were killed (reports differ) and about eighty injured. Damage was extensive[1–6]. The incident received considerable publicity, partly because an impressive set of colour photographs appeared in the French magazine *Paris Match*[6].

At the time there was no technical safety department in the Division but a senior member of the Technical Department, was asked to look into the implications. This work took up most of his time until he retired in 1970. Details of the fire were difficult to obtain from published sources and the main source of the information was a factory inspector who had visited the site.

Gradually, following discussions with the works, the engineering design department and other companies, new standards were established for new plants and proposals were made for improvements to old plants. Table 6.2 (see pages 78–80) is a summary of the main points, taken from a report issued in 1970.

These recommendations have stood the test of time well, though even today they are not followed by every company. Today, however, we would want to make the following changes and additions to the list:

- Flammable gas detectors should be installed where experience shows that leaks may occur.
- Vessels exposed to fire should be protected by depressuring as well as by water cooling and insulation[7] (item 6).
- In recent years the protection of LPG vessels by mounding (that is, by covering with clean sand or gravel) has become increasingly common. This makes a BLEVE impossible; water cooling is obviously not required and some of the other recommendations can be relaxed.
- Remotely operated emergency isolation valves are better than excess flow valves, as the latter can pass a large flow, up to twice the normal flow, without closing (item 14).
- Discharge of gas to atmosphere may be undesirable (or illegal) on environmental grounds even when it is safe.
- Most important of all, today we would (or should) start a safety study of a new or existing LPG installation by asking if we can reduce the amount of LPG in storage and process, or if we need use LPG at all (see Section 6.5, page 83). If large quantities (thousands of tonnes) must be stored we prefer refrigerated storage at low pressure rather than storage under pressure at ambient temperature.

Proposals were prepared for modifications to existing installations and considerable sums of money were spent. However, one works felt that only minor changes were necessary to their equipment. In 1970 there was a leak from a pump in its LPG storage area. Though the leak ignited, it was soon stopped and damage was slight, but it brought home to the works staff, far more vividly

TABLE 6.2
Safety in design of plants handling liquefied light hydrocarbons.
An extract from a 1970 report by H.G. Simpson.

Summary of the main points

Experience shows that the cloud of vapour generated from any major escape of liquefied hydrocarbon is likely to find a source of ignition, often with disastrous results.

Because of this the basic approach to safety in processing, storage and handling of liquefied hydrocarbons must be one of prevention, aiming to eliminate accidental escape wherever possible and to ensure that such spillage as does occur is restricted to a manageable quantity and can be dispersed safely.

Existing recommendations for safe practice in handling liquefied hydrocarbons conflict among themselves in a number of important aspects, creating the need for reassessment of the associated hazards and methods of dealing with them.

The following measures for safety in design of plants handling liquefied hydrocarbons are recommended for adoption by Petrochemicals Division.

1. Particular care should be taken in the selection of materials of construction of equipment which may at any time contain liquefied hydrocarbons at sub-zero temperatures.

2. The capacity of fire relief valves should be determined as described in American Petroleum Institute (API) Recommended Practice RP 2000 and not as described in API RP 520, as was formerly the practice.

3. Particular care should be taken with the arrangement of vent lines from relief valves on low pressure refrigerated storage tanks to avoid causing excessive back pressure.

4. Liquid relief valves should be provided on pipelines or other equipment which may be endangered by thermal expansion of locked in liquid.

5. Small branches on pressure vessels and major pipelines should be supported mechanically to prevent them from being broken off.

6. Particular care should be taken to protect equipment against fire exposure by a suitable combination of water cooling and fireproof insulation to ensure that metal temperatures cannot rise sufficiently to cause failure at relief valve set pressures.

7. No attempt should be made to extinguish a liquefied hydrocarbon fire except by cutting off the supply of hydrocarbon, nor should a cloud of vapour from an escape which is not on fire be deliberately ignited.

TABLE 6.2 (continued)
Safety in design of plants handling liquefied light hydrocarbons.
An extract from a 1970 report by H.G. Simpson.

8. Pumps should in general be fitted with mechanical seals instead of packed glands to reduce leakage.

9. Process draining and sampling facilities should be designed to withstand mechanical breakage, to minimise the risk of blockage by ice or hydrate and to restrict the quantity of any spillage. There should be a robust connection and first isolation on the plant or storage vessel and a second valve, of not more than ¾ inch (19 mm) size for draining or ¼ inch (6 mm) for sampling, separated from the first by at least 3 ft (1 m) of piping. The discharge pipe from the drain of not more than ¾ inch (19 mm) bore should deliver clear of the vessel and be supported to prevent breakage by jet forces. Both valves should have means of actuation which cannot be readily removed. Samples should be taken only into a bomb through a closed ¼ inch (6 mm) bore piping system.

10. Pressure storage vessels should preferably be designed with only one connection below the liquid level, fully welded up to a first remotely operated fire safe isolation valve located clear of the area of the tank.

11. Valve connections should be provided on process vessels for disposal of residues of liquefied hydrocarbons, preferably to a closed flare system. No bleed direct to atmosphere should be of more than ¾ inch (19 mm) bore.

12. Remotely controlled isolation valves should be provided on items of equipment which are liable to leak significantly in service.

13. Discharge of heavy vapour from relief valves and blowdowns should be vented to a closed system, preferably with a flarestack, except when it is possible to discharge to atmosphere at sufficient velocity to ensure safe dilution by jet mixing with air.

14. Excess flow valves should be installed in liquid and vapour connections which are regularly broken to atmosphere, particularly the flexible hose connections used in tank wagon operations.

15. Whenever possible, equipment should be located at the safe distances from sources of ignition determined by area classification. The horizontal extent of Division (now Zone) 2 areas in electrical classification should be taken the same as the safe distance.

TABLE 6.2 (continued)
Safety in design of plants handling liquefied light hydrocarbons.
An extract from a 1970 report by H.G. Simpson.

16. The ground under pressure storage vessels should be impervious and should slope so that any liquid spillage will flow away from the vessels to a catchment area where it can be safety disposed of, or can burn if it ignites without causing further hazard. Suitable diversion and retaining walls should be provided to prevent uncontrolled spread of the spillage. The height of the walls should be suitably limited in relation to their distance apart to allow minor leakage to be dispersed by natural air movement. The retention capacity for liquid should be decided in relation to the amount likely to escape allowing for flash-off and boil-off from the ground.

17. Low pressure refrigerated storage tanks should be fully bunded, and the floor of the bund should be sloped so that spillage flows preferentially away from the tank.

18. The principle of diverting liquid spillage away from equipment should be applied in process areas wherever possible.

19. In plant or storage areas where safe distances from sources of ignition cannot be met, or in areas near a factory perimeter adjacent to public roads or property the installation of a steam curtain should be considered.

For changes to this list see Section 6.3, page 77.

---

than the pictures from Feyzin, what an LPG fire is like. They made extensive modifications, resited pumps, removed redundant equipment and installed gas detectors and remotely operated isolation valves.

The springs of action in this section were the incidents, at Feyzin and locally, backed up by the detailed, carefully-reasoned arguments of one man. The Board was supportive, arranging for a senior member of the company to spend so much time on the problems and sanctioning the necessary expenditure.

The Factory Inspectorate was the main source of information on what had occurred at Feyzin and suggested some changes that might be made. These changes were not accepted uncritically, nor were they expected to be. This is the only one of the five changes discussed in this chapter in which the Factory Inspectorate was involved technically. Today its involvement would be much greater.

6.4 HIGH INTEGRITY PROTECTIVE SYSTEMS

This is another example of a change resulting from recognition of a problem, but this time the change was a novel design, not a new procedure.

During the 1960s the Petrochemicals Division of ICI designed and built two plants in which ethylene in the vapour phase was oxidised with oxygen; one was for the manufacture of vinyl acetate, the other for the manufacture of ethylene oxide. Both had to operate close to the explosive limit and it was obvious that a serious explosion could occur if the concentrations of the reactants departed even slightly from normal operating conditions. Reports of minor explosions had been received from other companies.

The Division debated the wisdom of constructing such potentially dangerous plants. It considered protection by blast walls but turned it down because construction of a wall strong enough to resist an explosion was found to be impracticable. The Instrument and Electrical Design Group was therefore asked to see if the plant could be made safe by protective instrumentation. It realised at once that:

- instrumentation can be designed to reduce the chance of an explosion to any desired level, but zero chance is approached asymptotically and can never be reached. (For a note on asymptotes, see Appendix 6.1 on page 92.) Therefore:
- it is necessary to define numerically the level of safety that is acceptable.

The first attempt to define this level was to say that working on the oxidation plants should be as safe as travelling by train. Later this was changed to say that it should be as safe as working on an average ICI plant. This change produced a small increase in the standard of safety.

To achieve the level of safety specified, the probability of an explosion in the reaction system had to be reduced to once in 30 000 years or less. The derivation of this figure is described elsewhere[8].

To achieve this level of safety fifty trips had to be installed. This would have produced many spurious trips and, to prevent these, each trip was made into a 2-out-of-3 voting system. Two measuring devices (trip initiators) out of three had to indicate an approach to a dangerous condition before a trip would operate. All the trips isolated the oxygen supply by closing three valves in the oxygen line and opening two bleed valves located between them, the whole system being duplicated so that it could be tested with the plant on line (see Figure 6.2 on page 82).

Early on it was realised that, with a system of this complexity, the design had to be checked in detail by a team independent of the designers.

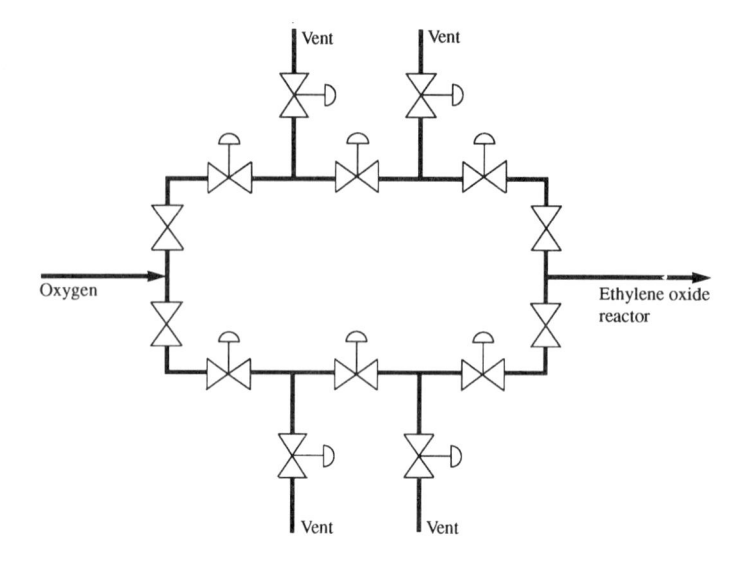

Figure 6.2   Arrangement of valves on oxygen inlet line to ethylene oxide reactor.

The whole system was more complex than that used by other ethylene oxide manufacturers, and turned out to be more complex than anyone expected it to be when the safety target was adopted. Nevertheless, the logic of the approach was accepted and the system sanctioned by the ICI Board. It has been described by Michael Stewart, the engineer responsible for the detailed design[9]. Probably no other chemical plant, to this day, contains a more complex protective system, though far more complex systems are installed on nuclear power plants.

Today the use of hazard analysis (also called hazan, risk analysis or probabilistic risk assessment) is commonplace. The authorities in many countries have encouraged or even insisted upon the use of numerical methods and the UK the Health and Safety Executive has published criteria for what it calls tolerable risk[10,11]. Sophisticated computer programs are now available for estimating risks to life and drawing risk contours round a plant. Stewart's design is still interesting, however, as the first complex chemical plant protective system to which the methods, developed in the nuclear industry, were applied.

The vinyl acetate plant had a short life but the ethylene oxide plant is, at the time of writing, still in operation and at least one other plant with a similar protective system has been built. Would we do the same if we were starting today?

Stewart's design assumed that if an explosion occurred an operator would be killed. Since then reports on 18 explosions on the reaction sections of ethylene oxide plants have become available and in only one case was someone killed[12]. It is possible that unreported deaths occurred in other cases, but even so it seems that Stewart's assumption was pessimistic. The reaction takes place in the vapour phase, inventories are therefore limited and any incident is a localised affair and does not devastate the site. Explosions in the distillation sections of ethylene oxide plants have been more serious.

The high integrity protective system solved a technical problem and it has worked successfully for many years. Nevertheless, in a way it was a second best solution. It was the supreme example of solving a safety problem by adding onto the plant a vast amount of expensive protective equipment, which requires extensive testing and maintenance and which could, in many companies, easily become neglected. The ideal is to see if, by redesign, we can remove the hazard, as I discuss in the next section, but I have to admit that in this case I cannot see a way of doing so.

6.5 INHERENTLY SAFER DESIGN — A SLOWER CHANGE

The four changes just described came about fairly quickly. The first, improvements in the permit-to-work system, followed a serious accident and was imposed immediately by decree. The second, improvements in trip testing, took place gradually over a period of several years as one group after another were persuaded to change their methods. The third, improvements in equipment for handling liquefied flammable gases, took place over a couple of years as financial sanction had to be obtained and equipment had to be designed and ordered. The fourth, development of a high integrity protective system, took place during the design of a new plant.

In contrast, a fifth change has been slow. Following the explosion at Flixborough in 1974 chemical engineers began to realise that the best way of preventing a leak of hazardous material is to use so little that it does not matter if it all leaks out or to use a safer material instead. I published a number of papers advocating this view during the late 1970s[13–15]. ('What you don't have, can't leak.') Although there has been much progress, industry has been much less willing to adopt these inherently safer designs, as they are called, than to adopt numerical methods of risk assessment, probably for the following reasons:

(1) A reluctance to admit that large inventories of hazardous materials are unsafe. The usual view has been that we know how to handle them safely, so there is no need to worry. When they have not been kept under control it is

blamed on the lower standards followed by other companies. When a company has had a leak on one of its own plants, it was an isolated incident that should not happen again.

Slowly, people are coming to realise that while equipment and people are on the whole very reliable, they are not reliable enough. When we are handling hazardous materials only very small failure rates, of equipment and people, are tolerable, and these low rates are difficult to achieve, year in, year out. We may keep up a tip-top performance for an hour or two, while playing a game or a piece of music, but we cannot do so all day, every day. We should therefore use designs that are tolerant of failure and error.

(2) Most inherently safer designs are difficult to backfit on an existing plant and the chemical industry has passed through a period of recession, with few new plants being built. However, intermediate stocks can be reduced on an existing plant, but little change was made until after Bhopal (see paragraph (5) below).

(3) Another possible reason is that the 1960s was a period of rapid change in the chemical industry which brought many problems in its wake. Many plants were difficult to start up, took a long time to achieve flowsheet, had to undergo expensive alterations and suffered many fires and explosions (see Section 7.4, page 96). The young managers who suffered the effects of innovations that were not thoroughly thought through are now senior managers, suspicious of innovation, who prefer to stick to the processes they know.

(4) Compared with the first four changes discussed, we are not asking for new equipment or new ways of carrying out specific tasks but for far-reaching changes in the design process. Most safety studies come late in design when all we can do is add onto the plant protective equipment to control hazards, equipment which may fail or be neglected. Inherently safer designs need systematic searches for alternatives in the first stages of design, at the flowsheet stage and even earlier at the conceptual stage when we decide which product to make, by what route and where to locate the plant. These searches amount to a cultural change, something which is always slow and which will not come about unless senior mangers get involved.

Those, like me, who have advocated the development of inherently safer designs, have had ready access to the people who were able to adopt the first four changes but less access to senior managers. Senior managers do not, on the whole, attend safety conferences or read safety journals and they often look on safety as something they exhort their staff to do better, rather than as an area in which they should be identifying the problems and monitoring progress (see Section 7.10, page 115).

(5) Company organisation has been another constraint on the development of inherently safer designs. Many companies are now organised by business rather than function; that is, instead of having departments for research, production, design, development and so on, all those concerned with a product or range of products are grouped together. With such an organisation it may be no-one's job to follow up innovations which benefit the company as a whole but do not produce an outstanding benefit for any one business. For example, the Higee process for carrying out distillation in rotating equipment[16] promises large economies in the long term for the major oil and chemical companies but very few have been built. Understandably, individual business or project managers are reluctant to install new designs in case there are any unforeseen problems, and there is nowadays no central department or budget which can underwrite novel equipment on behalf of the company as a whole.

Again, change will not come about unless senior managers become involved.

While this book was in production the technical press reported that the ICI research director has introduced a strategic R&D fund to support ICI scientists in projects outside the range of the individual businesses' strategies[27].

In 1984, ten years after Flixborough, a toxic gas release at Bhopal, India killed over 2000 people. The material which leaked, methyl isocyanate, was not a product or raw material but an intermediate, which it was convenient but not essential to store. Since then many companies have made a reduction of intermediate storage one of their aims[17], and they have also tried to reduce other stocks in storage. Progress at reducing inventories of hazardous materials in process has been slower.

What can be done to encourage the growth of inherently safer designs? All we can do, I think, is to continue patiently to advocate them to people at all levels, and hope that in time the ideas will be increasingly adopted.

6.6 OTHER SPRINGS OF ACTION

There is no doubt that a serious accident, especially one that achieves widespread publicity, is the most frequent cause of change in safety and the one that has the most rapid effects. However, as I have tried to show, some changes have come about because people recognised a problem and pressed for change. This section discusses briefly a few more influences on change.

TECHNICAL NOVELTY

If an incident illustrates novel and unusual technical features it is more likely to have an effect than a familar incident occurring again. This is partly because novelty appeals to the media and results in publicity and partly because companies are more willing to admit they have had an accident when it was technically novel and therefore hard to foresee. For example, in 1969 the vapour from a spillage of light hydrocarbons was sucked into the air inlet of a diesel engine and caused the engine to race. The driver tried to stop the engine by isolating the fuel supply but this had no effect as fuel was entering the engine through the air inlet; flashback occurred and ignited the vapour cloud. Two men were killed and the financial loss was large. The unusual source of ignition attracted considerable interest[18] but little attention was paid to the reasons why several tonnes of hydrocarbons were spilt.

Diesel engines had ignited clouds of flammable vapour before but the consequences had not been serious and the incidents got little publicity. The incident just described was the first to draw widespread attention to the fact that diesel engines can ignite flammable vapour just as easily as petrol engines, and so they should not be used in an uncontrolled way in areas where flammable gases and vapours are handled. It led to the development and marketing of devices for preventing ignition via the air inlet[19] and to the production of codes of practice for the safe use of diesel engines[20].

It will come as no surprise to readers to learn that diesel engines have continued to ignite leaks. For example, in 1984 petrol dripping from a loading arm caused the engine of a road tanker to race and emit black smoke. A fire could easily have started if the drips had not stopped[21]. The most spectacular incident of this type occurred in Russia in 1989 when a leak from a natural gas pipeline was ignited by a passing train, killing 650–800 people, according to press reports[22]. (It is not certain that the diesel engine was the source of ignition — sparks from the wheels or a match or cigarette thrown out of a window have been suggested — but the diesel engine seems the most likely source.)

INSURANCE COMPANIES AND THE LAW

Insurance companies and the law have helped to bring about much change but not innovation. This is not surprising, as until someone has invented a new device and shown that it works the insurance companies and the inspectorates can hardly insist on its use. The law follows good practice. It now follows it more quickly and energetically than in the past and this is welcomed by safety conscious companies which like to see their competitors made to adopt the safety measures already adopted in better companies (see Section 5.2, page 63).

Insurance companies have sponsored some research and in the UK the Health and Safety Executive has a large and very competent research division.

In 1988 the head of the Technology Division of the UK Health and Safety Executive wrote that 'The HSE may have been slow to exploit the full benefits of inherent safety, but I can assure you that ... we are now fully alert to its potential and will be following up inherent safety as it effects the design and operation of chemical plants'[23]. Though little has happened at the time of writing, there are signs that there will be more action in future[28].

PRESSURE GROUPS

These groups of concerned citizens are often activated by a desire to avoid development near their homes, the NIMBY (Not In My Backyard) syndrome, which can produce the BANANA (Build Absolutely Nothing Anywhere Near Anyone) effect. However, many pressure groups are made up of people with wider concerns who genuinely want a better world. They are a part of the democratic process but unfortunately often have a naive view of the relative extents of different risks and what it is and is not practicable to manage without. They can result in resources being given to those who shout the loudest rather than those with the greatest need or those who can use them to the best effect. So far as safety is concerned these pressure groups have been most active in the nuclear field and have resulted in large sums being spent to make small risks even smaller. Many more lives could have been saved if the money had been spent in other ways[8,24].

INDIVIDUALS

In one, extreme, view of history individuals do not count: if Napoleon or Hitler had not been born, some other Napoleon or Hitler would have arisen to do their work; their actions, it is said, were inevitable and the result of circumstances. There is a mite of truth in this view, in the sense that they succeeded only because the time was right, because there was a fertile field in which their ideas could take root, but without them history might have been very different. (See the quotation at the head of this chapter.)

Is change in safety the result of circumstances or individuals? The case studies in this chapter show that circumstances, such as a big incident, may make change inevitable, but individuals may determine the form that the change takes. When no big incident has occurred an individual can have an influence by recognising a problem earlier than others. To quote the German philosopher, Hegel (1770–1831), 'The great man of the age is the one who can put into words the will of his age, tell his age what its will is, and accomplish it ...'[25]. In Section 2.1, Incident 2 (page 6), for example, the new rules on the preparation of equipment for maintenance might have been less strict if the production director had been less strong-minded. (The same production director, Kenneth Gee, convinced the Board soon afterwards that an experienced technologist (me) should be appointed safety adviser, an unusual appointment at the time (see Chapter 9).)

6.7 HOW CAN WE ENCOURAGE CHANGE AND INNOVATION?

FIRST, WE HAVE TO CONVINCE PEOPLE THAT THERE IS A NEED FOR CHANGE

A major incident is the most effective persuader, particularly if it occurs near home (see Section 6.3, page 76). Sometimes the safety adviser has to persuade his colleagues not to over-react and spend too many resources on one problem.

The cases described in Sections 6.2 (page 74) and 6.4 (page 81) show that it is possible to convince people that change is necessary without waiting for a serious accident to occur, but the safety adviser then has a harder job. Conversion is slow, not sudden.

Note that in the cases described the emphasis was put on persuading people at the working level, not on persuading the Board to issue an edict, but this will be effective only if the company culture is favourable and the directors concerned are known to be broadly sympathetic.

SECOND, WE HAVE TO FIND A SOLUTION

This is a technical, not a human relations, problem and therefore the least difficult of those described. Sometimes, as in the decision to lock and blind, the solution is simple (though it turned out to be more complex than expected); sometimes, as in the decision to design a high integrity protective system, the solution is complex and has to be developed over a period of time.

THIRD, WE HAVE TO SELL OUR SOLUTION

As I have said, after an accident this is easy. If we have merely seen a problem, it is more difficult. Colleagues who have reluctantly admitted that there may be a problem may say that the solution is too expensive or impracticable, will interfere with production, introduce teething problems and may have unforeseen effects. Table 6.3 lists some of the arguments that may be met. With patience and persistence these objections can be overcome. The safety adviser has to use all the techniques of the salesman and the evangelist. Arguments must be put clearly, objections answered patiently. Above all we have to make sure that our product is good and we must not be discouraged when success is not immediate;

TABLE 6.3
Some of the excuses people give for not doing what the safety adviser wants them to do.

- 'The industry standards don't ask for it.'
- 'Our competitors don't do it.'
- 'We have been doing it this way for 20 years and never had an accident.'
- 'Why should we be an industry leader?'
- 'We can do it by changing the method of working without the need for new equipment.'
- 'I can't really believe in low probability numbers.'
- 'It's not my job.'
- 'I don't have the resources' and so on[26].

With such people, at least we know where we stand; we know we have to persuade them to do what we want, or talk to their bosses. A greater menace is the person who says, 'Yes. Certainly. No problem. I'll do what you want', and then does nothing. He also has a battery of excuses:

- 'I've been exceptionally busy.'
- 'We had a breakdown/major shutdown last month.'
- 'It's in next year's capital programme.'
- 'It's not a good time to ask the boss for the money.'
- 'We're looking for a suitable supplier.'
- 'My right-hand man just left.'
- 'I heard you had second thoughts about the project.'
- 'I thought we ought to get the safety committee's view.'
- 'We didn't have time to discuss it at the last design meeting.'
- 'I thought we might do it as part of the next revamp.'
- 'My boss isn't convinced it's a good thing.'
- 'There's a new code of practice out next year so I thought we ought to wait and see what is says.'

You can probably add some more.

---

it rarely is. Though the pace of change may be slower than we wish, it is fast by the standards of earlier generations.

Changes imposed by edict after a serious accident should be followed by persuasion or the change will not last (see Section 6.1, page 73).

FOURTH, WE HAVE TO MAKE SURE THAT OUR SOLUTIONS DO NOT LAPSE

This is the hardest part. Many people will enthusiastically sell or adopt a new idea and then lose interest. After a few years, blinding or trip testing lapses or new equipment falls into disuse. As I have already said, we have to keep reminding people of the reasons for our procedures, we have to check that they are being followed and we should not turn a blind eye when they are not.

REFERENCES IN CHAPTER 6

1. *The Engineer*, 25 March 1966, 475.
2. *Petroleum Times*, 21 January 1966, 132.
3. *Fire*, special supplement, February 1966.
4. Kletz, T.A., 1988, *What Went Wrong? — Case Histories of Process Plant Disasters*, 2nd edition (Gulf, Houston, Texas, USA), Section 8.1.1.
5. Lagadec, P., 1980, *Major Technological Risk* (Pergamon Press, Oxford, UK), 176.
6. *Paris Match*, 1966, No 875. Slides made from some of the photographs in the issue are included in *Hazard Workshop Module No 003, Fires and Explosions* (Institution of Chemical Engineers, Rugby, UK), undated.
7. Kletz, T.A., 1990, *Improving Chemical Industry Practices — A New Look at Old Myths of the Chemical Industry* (Hemisphere, New York, USA), 15.
8. Kletz, T.A., 1992, *Hazop and Hazan — Identifying and Assessing Process Industry Hazards*, 3rd edition (Institution of Chemical Engineers, Rugby, UK; published in the USA by Taylor & Francis, Bristol, PA), Sections 3.4 and 5.3.
9. Stewart, R.M., 1971, High integrity protective systems, *Symposium Series No 34* (Institution of Chemical Engineers, Rugby, UK).
10. Health and Safety Executive, 1992, *The Tolerability of Risks from Nuclear Power Stations*, 2nd edition (HMSO, London, UK).
11. Health and Safety Executive, 1989, *Risk Criteria for Land-use Planning in the Vicinity of Major Industrial Hazards* (HMSO, London, UK).
12. Kletz, T.A., 1988, *Plant/Operations Progress*, 7 (4): 226.
13. I advocated inherently safer design in a paper on the wider lessons of Flixborough, 'Preventing catastrophic accidents', *Chemical Engineering*, 12 April 1976, 83 (8): 124.
14. My first paper devoted entirely to inherently safer design was 'What you don't have, can't leak', *Chemistry and Industry*, 6 May 1978, 287.
15. I have discussed inherently safer design in more detail in *Plant Design for Safety — A User-Friendly Approach*, 1991 (Hemisphere, New York, USA).
16. Ramshaw, C, February 1983, *The Chemical Engineer*, No 389, 13.
17. See papers by D.E. Wade, D.C. Hendershot, R.J. Caputo and S.E. Dale in *Proceedings of the International Symposium on Preventing Major Chemical Accidents*, 1987, edited by J.L. Woodward (American Institute of Chemical Engineers, New York, USA).

18. *Chemical Age*, 12 December 1969, 40 and 9 January 1970, 11. Similar articles appeared in several other journals.
19. *Hazardous Cargo Bulletin*, January 1983, 21, 22, 24.
20. Oil Companies Materials Association, 1977, *Recommendations for the Protection of Diesel Engines Operating in Hazardous Areas*.
21. *Petroleum Review*, April 1984, 36.
22. *Daily Telegraph*, 5 and 6 June 1989.
23. Barrell, A., August 1988, *The Chemical Engineer*, No 451, 3.
24. Fremlin, J.H., 1989, *Power Production: What are the Risks?*, 2nd edition (Hilger, Bristol, UK).
25. Hegel, G.W.F., quoted by Carr, E.H., 1990, *What is History?*, 2nd edition (Penguin Books, London, UK), 54.
26. Ormsby, R.W., July 1990, *Plant/Operations Progress*, 9 (3): 166.
27. Stevenson, R., December 1992, *Chemistry in Britain*, 28 (12): 1071.
28. Jones, P., 10 September 1992, *The Chemical Engineer*, No 526, 29.

APPENDIX 6.1 — A NOTE ON ASYMPTOTES

Safety is often approached asymptotically. We wish to prevent a vessel being overpressured so we install a relief valve, properly sized and maintained. It may fail, so we install a second in parallel. The chance of failure is now much less but not zero, as both relief valves may fail at the same time. We could install a third and make the chance of failure even smaller, but still not zero. Where do we stop? To answer this question we have to set a target; we have to decide that overpressuring of the vessel is acceptable if it occurs less than once in a hundred years, or a thousand years or a million years. We cannot set a target of 'never' — unless we can remove the hazard by a change in design: making the vessel so strong that it will withstand any pressure to which it might be subjected. And even then, we cannot be sure that the design, construction, installation, operation and maintenance are up to standard.

Asymptotes can be illustrated by the story of the engineer who wooed a reluctant lady mathematician. She suggested that he stood some distance away and with each step halved the distance between them. As a mathematician she knew that they would never meet but as an engineer he knew that he would soon get near enough for all practical purposes. How near is 'near enough'?

7. THE MANAGEMENT OF SAFETY

'We judge ourselves by our policies. Others judge us by our actions.'
Anon

'Victory awaits those who have everything in order — people call that luck. Defeat awaits those who don't — this they call bad luck.'
Raold Amundsen (leader of the first expedition to reach the South Pole)[35]

This chapter discusses the more important factors that should be considered in the management of safety. (The term 'risk management' is usually given a more restricted meaning than 'safety management'. A risk manager is usually someone who decides which risks should be insured, which not, which risks should be accepted, which guarded against in other ways.)

7.1 POLICY

Many books and papers on the management of safety attach great importance to a formal statement of a company's safety policy. While such statements are necessary, and are required by law in the United Kingdom, I do not believe that they have much effect on a company's accident record, for several reasons:

- They are often mere vague statements of good intention of little practical use, sometimes written to satisfy the law enforcers.

- They sometimes give the impression that they are written to protect the writer rather than help the reader; ('Whoever is blamed for the accident it won't be me').

- They often exaggerate the priority actually given to safety, and everyone knows it.

- If employees ever read them, few can remember what they say. They are treated as so much 'noise', information of little value which is 'tuned out'.

- In theory, the directors of a company lay down the policy and the rest of us follow it. The practice is often different. We deal with problems as best we can, subject to various constraints. Looking back we see a common pattern. That is our policy. A good statement of policy, and one that will be followed, puts into the form of 'statute law' what is already the 'common law' of the company. Policy follows action more often than action follows policy. Policy statements which run counter to the 'common law', what is often called 'custom and practice', are likely to be ignored.

Sir John Harvey-Jones, later to become chairman of ICI, has described his experience in Intelligence[36]:

'I had always imagined the great men striding up and down their offices formulating in their infinitely wise minds the complex intersecting grand designs which would then be turned into action plans by serfs like me. Not for a moment had I dreamed that the leadership applied from the top took the form of rather negative interventions on the ideas flowing up from below ... progress was made more by the establishment of case law than anything else.'

If statements of policy do not affect actions, what does? The culture or 'common law' of a company is more important and this is conveyed by little things, such as a telephone call from the head office, immediately after an incident, asking not if anyone is hurt but when the plant will be back on line. Or just the expression on a manager's face when he is told that an operator shut the plant down because the water level in a boiler was low, but the alarm turned out to be false. If I want to know a company's policy on canteens, I do not ask for a statement of policy as a meal in the canteen will tell me more. It is the same with safety.

Changing a company's culture is much more difficult than issuing a new policy statement (see Section 6.7, page 88). Organisations are not putty in the hands of new managers; they are more like rubber: their traditional customs and practices constrain and push back.

7.2   WHOSE RESPONSIBILITY?

The designers of a new plant, the operating staff on an existing plant, are responsible for its safety, not the safety department. The experts in the safety department are there to advise, monitor, provide information, train and assist with studies such as hazard analyses and hazard and operability studies. They should say what they think should be done, and be prepared to take responsibility if the advice proves to be wrong. But they should not take away the designer's or manager's responsibilities or his right to say, 'Thank you for your advice, but I think I will do something different'.

Obviously the advice of an expert should not be rejected lightly, especially if his reputation is high, and especially if he is recommending a more cautious policy than the one the manager would like to follow. Experts, however,

THE MANAGEMENT OF SAFETY

do not always agree with each other and sometimes they see only part of the problem (see Section 9.1, page 143). It is easier for the manager to reject the expert's advice when the manager is the more cautious of the two. The expert, if he feels strongly enough about something, should be able to appeal to the manager's boss.

Safety experts should not, of course, wait until they are asked for advice. They should keep their ears to the ground, find out what is going on, and follow up; they should 'drop in' on plants which have had accidents and they should ask designers when they will be ready to discuss their new designs. They should be experts in communication as well as the technical problems of their industry. They should try to anticipate problems and not just react to accidents (see Sections 6.2 and 6.4, pages 74 and 81 respectively).

Because designers and line managers are ultimately responsible for safety, some companies have drawn the conclusion that there is no need to employ high level technical people in the safety department. All they need is a junior member of staff to number the dead and injured on the industrial battlefield. According to the official report on the fire at King's Cross Underground railway station in 1987, London Underground believed that passenger safety was 'inextricably intertwined with safe operating practices', and so did not employ specialist safety staff to advise line managers on passenger safety (but only to advise them on employee safety)[1]. The account of the fire, in which 31 people were killed, shows how much the line managers needed such advice (see Section 4.5.1, page 56).

Other companies may go to the opposite extreme and allow their managers to leave safety to the safety adviser. They may believe that if we follow all the rules we will have a safe plant and so the safety adviser's job is to know the rules. In the UK, the Health and Safety at Work Act (1974) requires companies to provide safe plant and methods of work and adequate instruction, training and supervision. It is not sufficient to follow the rules or even the codes (see Section 8.9, page 136).

In every organisation the line management forms a hierarchy or tree, and there is also a tree for each advisory or staff function such as accountancy or safety. At some point each staff tree merges with the line tree and becomes subordinate to it. This may be at a high level — most companies have a finance director on the Board — or a low level. If the senior safety adviser in a company is responsible to a junior or middle manager he will find it difficult to influence decisions. If he is on the Board, or directly responsible to a member of the Board, his task is easier. The attention paid to advice depends primarily on the ability of the adviser and the quality of his advice, but it also depends on his status[2].

7.3 TRAINING, INSTRUCTIONS AND LEARNING THE LESSONS OF THE PAST

These are the subjects of the other chapters in this book. The actions needed have been discussed already and are summarised in Chapter 10. I mention them here so that readers looking for a list of the topics covered by the management of safety do not overlook them.

As I said in the Introduction, it is not lack of knowledge that prevents our safety record being better than it is, but a failure to use the knowledge that is available, much of it acquired as the result of past accidents. If we do not know the action we should take, then others do. One of the major responsibilities of the safety professional is to continually remind his colleagues of the lessons of the past, using all the techniques of the professional communicator (see Chapter 10).

7.4 AUDITS AND THEIR LIMITATIONS

An audit is a systematic examination of an activity to see if it is being carried out correctly. Safety performance should be audited as well as accounts.

In 1970 the Chemical Industries Association carried out an investigation into safety auditing in the USA. Their report, *Safe and Sound*, which did much to introduce the practice to the UK, reminded us that ' ... the standard of performance of any given function varies directly with the degree of probability that someone higher up is going to look at what has been done and what has not been done'. Hence managers at all levels should carry out a certain amount of auditing as part of their job. They may wish to set aside special periods of time, or they may simply keep their eyes open as they go round the plant on their normal visits. It does not matter, so long as they do it.

In addition, more formal audits should also be carried out by a team or an individual from outside the plant who spends a few days or even a few weeks on the job. A typical team might include a professional safety auditor, someone from another plant, a safety adviser and an expert on the hazards involved. Inviting people from other companies to take part in audits has been suggested but not welcomed. In addition, factory inspectors and insurance surveyors may carry out their own audits.

We need these audits by outsiders because:

- Those who work in a plant do not notice the hazards they see every day; vibrating pipes become part of the background. But they may be noticed by an outsider.

- Auditors may have specialised knowledge and thus see hazards not apparent to others.
- Auditors have more time for investigation in depth than those who work regularly on a plant.

Thus safety auditing should not be a police activity — to catch people who are not doing their job properly — but an activity designed to help the local management who may miss hazards through familiarity, ignorance or lack of time.

Many auditors look only for the obvious mechanical hazards and ignore methods of working, or software as they are often called. As well as mechanical hazards, auditors should also look at:

- The quality of the training and instructions (see Chapter 10).
- The procedures for preparing equipment for maintenance and controlling modifications (see Section 7.8, page 109) and testing protective equipment (see Section 7.5, page 101) and whether or not these procedures are actually followed.
- Procedures for investigating accidents, passing on the lessons learned and ensuring they are not forgotten (see Section 7.7, page 106).
- Process hazards as well as mechanical ones.
- The quality of the management and the effectiveness of their commitment to safety. Vague protestations of good will are not enough (see Section 7.10, page 115).
- Places which others do not look at — behind and underneath equipment, at the backs of buildings, structures reached only by catladders. The main production area may appear fine but an examination of these other places may disclose a different state of affairs. Rot starts at the edges.
- Although it may be a separate exercise, process hazards should be reassessed every few years in the light of new knowledge and any new techniques that have become available (see Section 8.3, page 127).

Auditors (and managers) should visit the plant at night and at weekends, not just during the day.

Time is always limited. Instead of trying to have a quick look at everything it may be better to pick one topic at a time and consider it in depth. Such selective audits are usually known as surveys. If we wish to excavate a field for archaeological remains, we could remove an equal depth of soil from the whole field or we could dig out a small area down to the underlying rock. *If the area is well-chosen* the latter will produce more results, especially if resources are limited, as they usually are.

Table 7.1 shows some of the surveys that were carried out during my period as a safety adviser with ICI Petrochemicals Division, mainly by people who spent most of their time on surveys and became expert in the subjects surveyed. The reasons why these subjects were chosen are shown.

Note that in a survey we try to look at every item in the area selected, not just a sample. When tests of trips and alarms were surveyed, we witnessed the testing of every trip and alarm in the Division (see Section 6.2, page 74), at the time an organisation employing about 10 000 people. When sampling was surveyed every sample point was inspected and most were seen in use. But when

TABLE 7.1
A list of surveys carried out in ICI Petrochemicals Division and the reasons why these subjects were chosen.

Subject of survey	Reason why chosen
Equipment:	
Testing of alarms and trips	Recognition of the importance of regular testing (see Section 6.2, page 74)
Classified electrical equipment	Comments by a factory inspector following an explosion
Sampling	Several accidents
Flame traps and open vents	Several accidents including one fatal
Equipment for handling liquefied flammable gases	The Feyzin fire (see Section 6.3, page 76)
Foam-overs	A serious incident overseas and lesser incidents at home
Level glasses	Serious incidents elsewhere
Flammable dust hazards	Reorganisation brought them into the Division
Procedures:	
Permits-to-work	A serious fire due to a poor permit system (see Section 6.1, page 70)
Modification control	Flixborough
Registration and testing of relief devices	Recognition of the importance of regular testing
Testing of gas detectors	Realisation that they do not always fail safe

flameproof electrical equipment was surveyed, we inspected only every tenth item, while every hundredth was dismantled in the presence of the surveyor.

If we can quantify the results of an audit we can see if standards have changed since the last audit and how one factory or company compares with another. ('What gets measured gets done' — Peter Drucker.) Several methods of quantifying audits results have therefore been devised, of which the best-known is the International Safety Rating System[3]. Marks are awarded under a large number of headings and combined into an overall index. For example, in the South African version of the scheme — the 5-Star Safety System — portable electric equipment can get up to 30 points, of which up to 20 are given for the condition of cords, plugs and switches and up to 10 if the equipment is identified, checked and a register kept. Up to 40 points can be given for the quality of injury reporting and investigation (under five sub-headings) and another 40 for the quality of damage reporting and investigation.

The International Safety Rating System does not cover the specific technical hazards of a plant which must be assessed separately, something which is not always realised. BP have developed a Loss Exposure and Technical Safety Audit (LETSA) for use alongside the International Safety Rating System. In auditing pumps handling highly flammable or toxic liquids, for example, negative marks are given if the pumps are located beneath cables, flare lines or other equipment, and positive marks if they are fitted with emergency isolation valves and high integrity seals and gas detectors are installed[4].

The Liquefied Petroleum Gas Safety Association of South Africa has developed a simple audit scheme suitable for small LPG installations[5]. It is based on the 5-Star system but is simpler and includes several questions on the hazards specific to LPG.

A weakness of many audits is one they share with all techniques which use a check list of subjects to be covered. Hazards not on the list are ignored, especially those new ones which creep up on us unobserved and take us by surprise. It is difficult to see their signal above the noise when we do not know what signal to look for, and are hoping there is no signal at all (see Section 8.1, page 125). Hazard and operability studies (Section 7.6, page 102) can help us see the hazards in new designs but only if someone in the team has met these hazards, or others like them, before. It is particularly difficult to detect gradual change, as the following examples show:

- Section 6.1 (see page 70) described how three men were killed and a plant was destroyed when a fitter opened up a 14 inch line and found it full of hot oil. On other occasions people had broken into lines which had not been properly isolated, but when lines were small they got away with it. As lines became bigger (and temperatures and pressures higher) more stringent methods of preparing equipment for maintenance became necessary, but no-one realised this until after the fire.

Similarly, in 1914 the nations of Europe did not realise how powerful the destructive powers of their armies had become. Kenneth Davis writes[6]:

'Perhaps the men raised in the nineteenth century on chivalrous, aristocratic ideals and wars still fought on horseback had no idea what havoc their twentieth-century arsenals could wreak. The world of sabres and cavalry charges was giving way to such inventions as mustard gas, U-boats, and the flamethrower (perfected by Germany), the tank (perfected by the British), and a new generation of hand-grenades and water-cooled machine guns.'

- Over the years, as a result of falling demand and plant closures, steam consumption in a factory gradually fell. The condition of some steam traps was poor; one was not working and another was isolated. This did not matter when steam flow was large as the flow moved the condensate along to another part of the main but when the flow fell condensate started to accumulate and ultimately water hammer fractured the main.

Another gradual change is described in Section 7.8 on page 112.

No index that I know of will help us measure the hazards we have not recognised but the following actions might help us see them:
- Accept that there probably is a signal, perhaps more than one.
- If you cannot see it, outsiders may, so include them in your audit teams.
- Do not let your audits become too structured. Always look at some things that are NOT on the list.
- Do not rely on the audit team to do your job for you. Get out and look for yourself.
- Pay particular attention to the areas where experience shows things are most likely to go wrong — for example, remote plants and peripheral operations which get less attention than the company heartland, service areas, transport operations and new acquisitions. Also, pay attention to times when things are most likely to go wrong, such as periods of rapid expansion or falling sales.

ICI pays immense attention to safety and has a good record but nevertheless an explosion occurred in a van carrying explosives because some fuseheads were packed in a rusty metal box. Rust greatly increases their sensitivity to friction[7].
- The theme of this book: coming events cast their shadows before them. After a serious accident we often learn that something similar has happened before

but no-one was hurt and damage was slight and so no-one did very much about it. Or an earlier accident was serious but occurred a long time ago and has been forgotten (see Section 2.1, page 4). Report and investigate near-misses and do not forget the lessons of the past.

7.5 TESTING AND INSPECTION

As discussed in Sections 3.2 and 6.2 (pages 29 and 74 respectively), all protective equipment, and all equipment on which the safety of the plant depends, should be tested or inspected regularly to check that it is in working order and is fit for use. If we do not do so, the equipment may not work when required. The frequency of testing or inspection depends on the failure rate. Relief valves are very reliable; they fail about once per hundred years on average, and testing every one or two years is usually adequate. Protective systems based on instruments, such as trips and alarms, fail more often — about once every couple of years on average — so more frequent testing is necessary, about once per month. Pressure systems (vessels and pipework) on non-corrosive duties can go for many years between inspections but on some duties they may have to be inspected annually, or even more often.

The following sorts of protective equipment should also be tested or inspected regularly, though they are often overlooked even by companies which conscienciously test their relief valves. In some cases, times of response should be checked.

- Drain holes in relief valve tailpipes. If they choke, rainwater will accumulate in the tailpipe.
- Drain valves in tank bunds. If they are left open the bund is useless.
- Emergency equipment such as diesel-driven fire water pumps and generators.
- Earth connections, especially the moveable ones used for earthing road tankers (see Section 7.7, page 108).
- Fire and smoke detectors and fire-fighting equipment.
- Flame arrestors.
- Hired equipment. Who tests it — the owner or the hirer?
- Labels are a sort of protective equipment. They vanish with remarkable speed, and it is worth checking regularly to make sure that they are still there.
- Mechanical protective equipment such as overspeed trips.
- Nitrogen blanketing (on tanks, stacks and centrifuges).
- Non-return valves and other backflow prevention devices, if their failure can affect the safety of the plant.

- Open vents. These are in effect relief devices of the simplest possible kind, and should be treated with the same respect.
- Passive protective equipment such as insulation. If 10% of the fire insulation on a vessel is missing, the rest is useless.
- Spare pumps, especially those fitted with auto-starts.
- Steam traps.
- Trace heating (steam or electrical).
- Valves, remotely operated and hand-operated, which have to be used in an emergency.
- Ventilation equipment.
- Water sprays and steam curtains.

All protective equipment should be designed so that it can be tested or inspected, and access should be provided. Audits should include a check that the tests are carried out and the results acted on.

The manager or engineer responsible should be reminded when a test or inspection is due, and senior managers should be informed if it has not been carried out by the due date. Test and inspection schedules should include guidance on the methods to be used and the features that should be inspected.

Test results should be displayed for all to see — for example, on a board in the control room.

Tests should be like 'real life'. In the example described in Section 3.2 (page 29), a high temperature trip failed to work despite regular testing. It was removed from its case before testing so the test did not disclose that the pointer rubbed against the case. This prevented it indicating a high temperature.

Operators sometimes regard tests and inspections as a nuisance, interfering with the smooth operation of the plant. Operator training should emphasise that protective equipment is there for their protection and they should 'own' it.

7.6   IDENTIFYING AND ASSESSING HAZARDS

Audits and surveys (Section 7.4, page 96) can identify the hazards on an existing plant, but other methods are needed during design. Check lists are sometimes used; we tick off the hazards that are relevant and cross off those that are not. The weakness of check lists is that, as already stated, new hazards not on the list are not brought forward for consideration and are therefore missed. Check lists are satisfactory if we are repeating a design we have used before, but not when we are innovating. Hazard and operability studies (hazops) are therefore preferred in the process industries. In a hazop each line in a line diagram is examined in turn by a team of designers and the commissioning manager under an independent chairman. The team members ask if no flow or reverse flow could occur in the line under examination. If it could, they ask if it would be hazardous or prevent efficient operation and, if so, what changes in design or method of operation would overcome the hazard or operating problem. The team then applies similar questioning to more and less flow, temperature, pressure and any other important parameters. Finally, they ask about the effect of changes in concentration or the presence of additional materials or phases. These questions should be asked about all lines, including steam, water, drains and other service lines, for all modes of operation, start-up, shut-down and preparation for maintenance, as well as normal operation[8].

Hazop is a powerful technique for identifying potential problems but is usually carried out too late for the team to *avoid* the hazards and all they can do is *control* them by *adding on* protective equipment. Similar studies should also be carried out much earlier in design — at the flowsheet stage, and even earlier at the conceptual stage — when companies decide which product to make, by what route, and where to locate the plant. Unfortunately many companies who carry out hazops late in design do not carry out these earlier studies. Their safety advisers do not get involved, their safety studies do not take place, until late in design and safety then becomes an expensive (though necessary) addition to capital cost. If they carried out the earlier studies they would be able, in many cases, to design plants that are both cheaper and safer[9] (see Section 6.5, page 83).

Having identified the hazards on a new or existing plant we have to decide how far to go in removing them (or protecting people from the consequences). We cannot remove all hazards, however trivial or unlikely to occur, and in the UK the law does not require us to do so. We have to do only what is 'reasonably practicable', weighing in the balance the size of a risk and the cost, in money, time and trouble, of removing it. In the process industries hazard analysis (also called hazan, risk analysis, quantitative risk assessment [QRA] or probabilistic risk assessment [PRA]) has been widely used to help us decide whether a risk is so large that it should be reduced, or so small that it can be ignored, at least for the time being[8].

In applying hazard analysis we have to answer three questions:

• How often will an accident occur? Experience can sometimes tell us the answer but often there is no experience and we have to estimate an answer from the known failure rates of the components of the system, using fault tree analysis.

• How big will the consequences be, to employees, to members of the public and to the plant? Again, whenever possible experience should tell us the answer but often there is no experience and we have to use synthetic methods.

- Finally, we compare the answers to the first two questions with a target or criterion. Various criteria, usually based on the risk to life, have been proposed and recently the UK Health and Safety Executive has made proposals[10].

Hazop is a technique that can (and should) be used on all new designs. Hazan, on the other hand, is a selective technique. There is no need, and we do not have the resources, to quantify every hazard on every plant. Having identified a hazard our experience or a code of practice usually tells us how far we should go in removing it. Hazan should be used only when the case for and against action is finely balanced and there is no experience or code of practice to guide us.

ICI has devised a 6-stage programme of hazard studies for identifying and assessing the hazards on process plants[11].

There are, of course, other methods of identifying and assessing hazards besides hazop and hazan and some may be more suitable for other industries. The nuclear industry, for example, often uses failure mode and effect analysis (FMEA). (This is the opposite of a fault tree. In drawing a fault tree we start with an event such as a fire and write down the various subsidiary events events that can lead up to the fire — for example, a spillage of flammable liquid, due to overflow of a tank, due in turn to failure of a high level alarm. In FMEA we start with the failure of an item of equipment such as high level alarm and write down the possible consequences such as overflow of the tank, a spillage and a fire.) It is not essential to use hazop or to use quantitative methods of assessing hazards. It is a management responsibility, however, to see that there is a systematic procedure for identifying and assessing hazards, suitable for the technology of the company, and that it is understood and used.

The process industries managed without hazop and hazan for many years. Why do we need them now?

We need hazop, or a similar systematic technique for identifying hazards, because plants have got bigger. The traditional way of finding out what could go wrong was to build a plant and see what happened; the 'dog is allowed one bite' method. Until our dog has bitten someone we do not have to muzzle it or keep it chained up. This method was acceptable when the consequences of a bite were small but not now that we have dogs as big as Bhopal, Piper Alpha and Flixborough.

We need hazan in part because society has become less tolerant of accidents and pollution but also because much equipment is less reliable than it used to be. While I was writing this book I visited Wigan Pier and saw the two magnificent preserved 1907 steam engines (Figure 7.1) which once supplied all

THE MANAGEMENT OF SAFETY

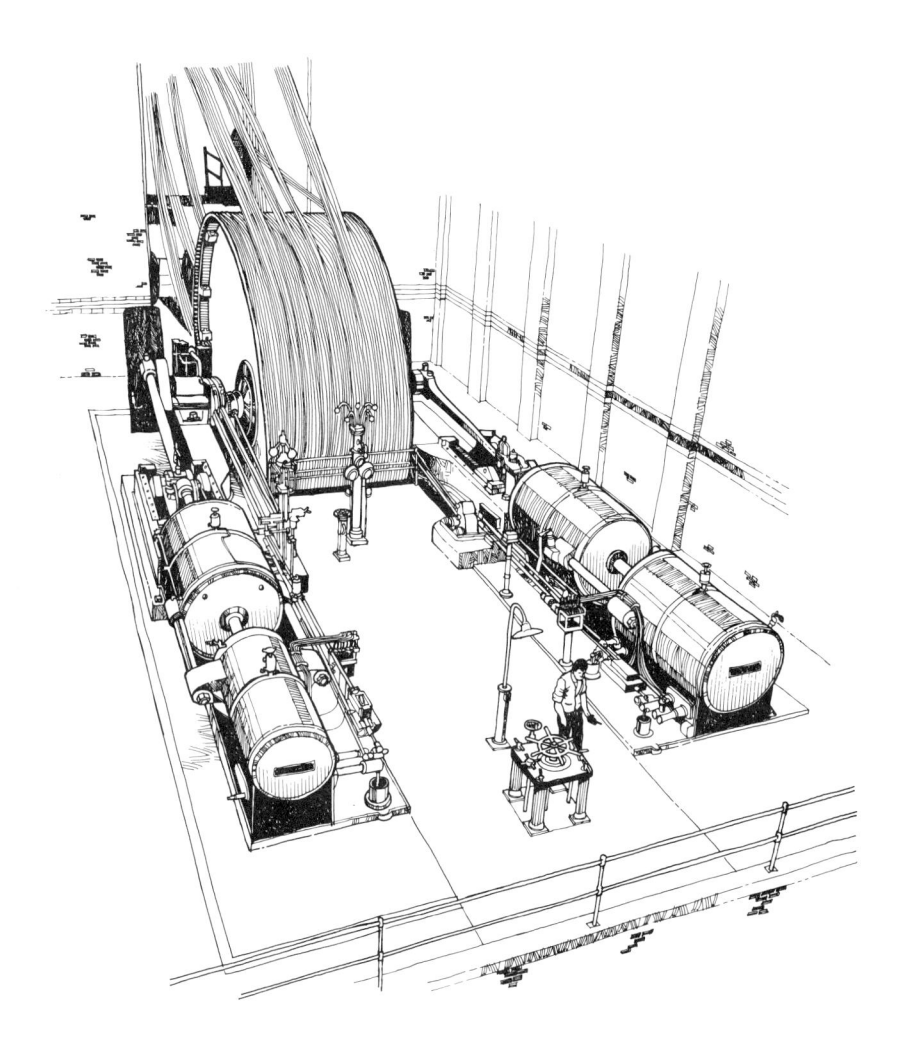

Figure 7.1 Trencherfield cotton mill, Wigan Pier. With these two steam engines and six boilers, the reliability of the power supply was not a problem.

the power needed by Trencherfield cotton mill. It struck me that with two steam engines, built like battleships, and six boilers the operators never had to worry about the reliability of their power supply.

Similarly relief valves have a low failure rate, typically once in 100 years, and few people worry about their reliability (unless the materials handled are dirty or corrosive or liable to freeze) but protective systems based on instruments are less reliable and we need to know how reliable they are.

105

7.7 ACCIDENT INVESTIGATION AND HUMAN ERROR

This book has emphasised the importance of learning from the past. Today is the past of the future. We and our successors can learn from today's accidents only if we investigate them, report them widely (see Chapter 5) and make sure that their messages are not lost. We should investigate near-misses as well as accidents, and accidents that caused only damage as well as those that caused injury. As previous chapters have shown, accidents which have minor results the first time they occur may kill people the next time.

The following are some points often overlooked in the first stage of an accident investigation, collecting the facts:

- Do not disturb evidence that may be useful to experts that are called in later, and photograph any equipment that has to be moved.

- Draw up a list of everyone that may be able to help, such as witnesses, workers on other shifts, designers and technical experts.

- Make it clear that the objective of the investigation is to find out the facts so that we can take action to prevent the accident happening again, not to find culprits who can be blamed. If people think that they, or one of their colleagues, will be blamed they may suppress some of the facts. An indulgent attitude towards people who have had lapses of attention, made errors of judgement or have not always followed the rules (see later) is a price worth paying to find out what happened (see Section 5.4, page 67).

- Be patient when questioning witnesses. Valuable background information will not come to light if we try to extract police-type statements but it may appear if we let people ramble on in as relaxed a manner as possible. Avoid questions to which the answer is 'Yes' or 'No' as it is easier for people to say 'Yes' or 'No' than to give long explanations, especially if they are suffering from shock. Remember that, while most people answer questions truthfully, they may not volunteer information so we may have to probe for it.

- Record information, quantitative if possible, on damage and injuries, amount of material involved and so on, so that others can use it for prediction.

- Notify any authorities that have to be informed and the insurance company, if claims are expected.

After we have collected the facts, we must draw conclusions and recommendations from them. These are not always obvious. As we have seen in earlier chapters, we should look beyond the obvious causes for the underlying causes. Accident investigation is like peeling an onion or dismantling a Russian doll. The outer layers deal with the immediate technical causes and triggering events, while the inner layers deal with ways of avoiding the hazard and with the underlying weaknesses in the management system. Reference 12 analyses a number of accidents, including major incidents such as Flixborough and Three Mile Island, in gradually increasing depth and shows how this can be done.

Dealing with the immediate technical causes of a leak, for example, will prevent another leak for the same reason. If we can use a safer material instead (or so little of the hazardous one that leaks do not matter) we prevent all leaks of this hazardous material. If we can improve the management system we may be able to prevent many more accidents.

The following are some other points to watch when drawing conclusions from the facts (see also the last part of Section 10.3 (page 171)):

(1) Avoid the temptation to list causes we can do little or nothing about. For example, a source of ignition should not be listed as the primary cause of a fire or explosion, as leaks of flammable gases are liable to ignite even though we remove known sources of ignition (see Section 2.2, page 10). The cause is whatever led to the formation of a flammable mixture of gas or vapour and air. (Removal of known sources of ignition should, however, be included in the recommendations.)

Similarly, human error should not be listed as a cause. See item (6) below.

Listing sources of ignition or human error as the cause of an accident is about as helpful as listing gravity as the cause of a fall. It may be true but it does not lead to constructive action.

(2) Do not produce a long list of recommendations without any indication of the relative contributions they will make to the reduction of risk or any comparison of costs and benefits. Resources are limited and the more we spend on reducing one hazard the less there is left to spend on reducing others.

(3) Avoid the temptation to overreact after an accident and install an excessive amount of protective equipment or complex procedures which are unlikely to be followed after a few years have elapsed. Sometimes an accident occurs because the protective equipment available was not used, but nevertheless the report recommends installation of more protective equipment; or an accident occurs because complex procedures were not followed and the report recommends extra procedures. It would be better to find out why the original equipment was not used or the original procedures were not followed.

(4) Remember that few, if any, accidents have a simple cause. As the accidents discussed in earlier chapters show there are usually many causes and many methods of prevention. We like to classify everything into tidy categories, and

allocate an accident to human error or equipment failure, but life does not cooperate; every accident has many causes. Bill Doyle, one of the pioneers of loss prevention, used to say that for every complex problem there was at least one simple, plausible, wrong solution[13].

(5) When reading an accident report, look for the things that are not said. The white spaces between the words may tell us as much as the words themselves. There is a story that to read a communist newspaper you first read what was written; then you tried to work out what really happened and what they were trying to make you think. I am not suggesting that the writers of accident reports deliberately falsify but, being human, they consciously or unconsciously present themselves and their companies in as good a light as possible (see Section 8.12, page 138). Try to look beyond the printed words.

For example, a gland leak on a liquefied flammable gas pump caught fire and caused considerable damage. The manager's report rightly drew attention to the congested lay-out, the amount of redundant equipment in the area, the fact that a gearbox casing had been made of aluminium, which melted, and several other unsatisfactory features. It did not stress that there had been a number of gland leaks on this pump over the years, that reliable glands are available for liquefied gases at ambient temperatures and therefore there was no need to have tolerated a leaky pump on this duty.

Another example: a fire was said to have been caused by lightning. The report admitted that the earthing was faulty but did not say when it was last checked, if it was scheduled for regular inspection, if there was a specification for the resistance to earth, if employees understood the need for good earthing and so on.

Appendix 7.2 on page 122 describes another failure to look for underlying causes.

(6) At one time most accidents were said to be due to human error, and in a sense they all are. If someone — designer, manager, supervisor, operator or maintenance worker — had done something differently the accident would not have occurred. However, when someone says that an accident was due to human error he usually means an error by someone at the bottom of the organisation who could not blame someone below him. We do not want to blame ourselves and it is easier to blame those below us than those above us. (I have seen this called 'organisational pressure to limit the boundaries of responsibility'.) This view is still found but responsible companies now accept that all accidents, including human errors, are management failures. To see how managers could prevent them we have to look more closely at human errors and not treat then as if they were all the same[14].

- Some errors are due to poor training or instructions — someone did not know what to do. It is a management responsibility to provide good training and instructions, and avoid instructions that are designed to protect the writer rather than help the reader. However many instructions we write, problems will arise that are not covered and so people, particularly operators, should be trained in flexibility — that is, the ability to diagnose and handle unforeseen situations.

If the instructions are hard to follow, can the job be simplified?

- Some accidents occur because someone knows what to do but makes a deliberate decision not to do it. We should:

— If possible, simplify the job. If the correct method is difficult, an incorrect method will be used.

— Explain the reasons for the instructions.

— Carry out checks from time to time to see that instructions are being followed and not turn a blind eye if they are not. Accidents rarely occur the first time someone fails to follow instructions; they usually occur after this has been going on for months or years and good management could have stopped it. (How often have you broken the speed limits on the roads before you were stopped?)

- Some accidents occur because the job is beyond the physical or mental ability of the person asked to do it, sometimes beyond anyone's ability. We should improve the plant design or method of working.

- The fourth category is the commonest: a momentary slip or lapse of attention. They happen to everyone from time to time and cannot be prevented by telling people to be more careful or telling them to keep their minds on the job. All we can do is to change the plant design or method of working so as to remove opportunities for error (or minimise the consequences or provide opportunities for recovery). We should, whenever possible, design user-friendly plants which can withstand errors (and equipment failures) without serious effects on safety (and output and efficency)[9].

Analysing human errors in this way does not remove responsibility, but it helps us distinguish between factors which are and are not within a person's power to control.

7.8 OTHER KEY PROCEDURES

When I retired from industry one of my first tasks was to sort a large number of accident reports that I had collected over the years from many companies. The thickest of the folders I produced contained reports on accidents which had occurred because the preparation of equipment for maintenance was poor. Sometimes procedures were poor, sometimes they were not followed. Some of

LESSONS FROM DISASTER

these accidents, and some more recent ones, have been described in Sections 2.1. and 6.1 (pages 4 and 70 respectively). The next thickest folders contained reports on accidents due to human error (see Section 7.7, page 108) and the unforeseen effects of modifications to plant or process. Sometimes there was no procedure for controlling modifications; sometimes the procedure was not followed.

Procedures for the preparation of equipment for maintenance and the control of modifications cannot therefore be taken for granted. Managers should satisfy themselves that their procedures are sound and are actually followed. As with all procedures, we should develop a good procedure, persuade people to follow it and check up to see that they are doing so. To emphasise this point, I shall describe a few accidents that occurred because these procedures were poor or neglected. Others are described in References 15 (preparation for maintenance) and 16 and 17 (modifications).

The incidents described in Sections 2.1 and 6.1 (pages 4 and 70 respectively) occurred because equipment was not adequately isolated from sources of danger. Other aspects of preparation for maintenance that are often neglected are identification of the equipment to be maintained and freeing from hazardous materials.

IDENTIFICATION OF EQUIPMENT FOR MAINTENANCE
- The following incident is typical of many.

Two reactors, about 2 m apart, and almost identical in appearance, were both fitted with radioactive level indicators. The reactors had different numbers but were usually known by their names which were rather similar. We will call them the polysolvent and bysolvent reactors. Neither names nor numbers were clearly marked on the vessels.

Both level indicators were removed while the vessels were repaired. The work on the polysolvent reactor was finished first and the Instrument Section was asked to replace the level indicator. The artificer installed it in error on the bysolvent reactor.

When the polysolvent reactor was brought on line the operators found that the level indicator was not working. They asked the Instrument Section to investigate and the error came to light. Fortunately it was a weekend and no-one was working on the bysolvent reactor.

Similar incidents have occurred many times before (see Section 3.6, page 33) and the message is clear: *All equipment prepared for maintenance should be clearly labelled and the name and/or number written on the permit-to-work. If the equipment has no permanent name or number a numbered tag should be attached to the equipment.*

- Here is another frequent scenario. Contractors were hired to demolish an old plant. They were told that they could burn the mild steel service lines but not the stainless steel process lines as they might contain product residues. To make it clear which were which, the service lines were tagged. The contractors found a line which was not tagged but was obviously mild steel so they started to burn it. It was a stainless steel process line inside a mild steel jacket. Residues in the pipe caught fire and the burner was severely injured.

If the plant could not be cleaned before demolition, the contractors should have been supervised more closely. They do not as a rule understand chemical plant hazards.

FREEING EQUIPMENT FROM HAZARDOUS MATERIAL

- The hold of a small ship was painted with a protective coating. The hold was ventilated while the coating was being applied and the concentration of solvent vapour was kept at a safe level. About six hours after the contractor had finished, a welder was asked to remove a steel stiffener from the outside of the hold. When he started work an explosion seriously damaged the ship. The coating had continued to give off solvent vapour for some time after it was applied and the welding ignited the vapour. Ventilation should have been continued for some hours after the contractor had finished and tests should have been carried out inside the hold before welding was allowed on the outside[18].
- Two men were trying to remove the manhole cover from a road tanker which had carried petrol and still contained a mixture of petrol vapour and air. To loosen the retaining bolts they apparently used sufficient force to produce a spark and the tanker exploded, killing both men[19]. This incident shows once again how easily mixtures of flammable vapour and air can ignite, and that we should never allow people to work where their ignition could cause injury (see Section 2.2 and Appendix 7.3, pages 10 and 124 respectively).

UNFORESEEN EFFECTS OF MODIFICATIONS

- Brunel's ship, *Great Britain*, completed in 1846, was the first iron, screw-propelled, ocean-going steamship ever built and at the time the largest ship ever built. On its maiden voyage from Liverpool to the Isle of Man it missed the island and ran aground on the coast of Northern Ireland. No-one foresaw that the iron in the ship would affect the compass. (No accident has a single cause and there was also an error in the charts.) Once the hazard was recognised it was easy to avoid it by mounting the compass on the mast and viewing it through a periscope[20].

LESSONS FROM DISASTER

- Eleven men were killed in an explosion in a steelworks in Scunthorpe, UK, in 1975. Molten iron at 1500°C was allowed to run out of a blast furnace into a closed ladle or torpedo — a long, thin vessel designed to carry 175 tonnes of iron and mounted on a railway chassis. The iron entered through a hole in the top, 2 feet (0.6 m) in diameter. Between two and three tonnes of water from a leak also entered the torpedo and rested on top of the molten iron, separated from it by a crust of iron and slag. When the torpedo was moved the water came into close contact with the molten iron, vaporised instantaneously with explosive violence, and blew out 90 tonnes of molten iron.

    A flow of water onto the top of molten iron in an old-fashioned open ladle was not a hazard as there was plenty of room for the steam to escape. When the design of the ladle was changed no-one realised that the entry of water was now dangerous. It is not difficult to prevent the entry of water once we see the need to do so.

    The official report[21] described the reason for the water leak in detail but this is of little importance. The underlying cause of the explosion was the failure to foresee the effects of the change in design and that, in turn, was due to the lack of any procedure for studying the possible effects of change. This was not uncommon in industry at the time. The explosion at Flixborough[22,23] in 1974, the result of a temporary modification, encouraged many companies to introduce formal procedures for examining modifications.

- An example of an accident caused by a change in procedures: an operator was asked to add several reactants to a pilot plant reactor. His experience told him that there would probably be a fume emission so he changed the order of addition. This caused a runaway reaction.

    The operator should not have made the change without the authorisation of the chemist in charge. But equally, if the chemist had discussed the procedure with the operator, he would have learnt about the fume problem[24].

- Two gradual changes were described in Section 7.4 (see page 100). Here is another. The canning of food began in 1812 with the Admiralty as the main customer. Some time after 1845 an outbreak of food poisoning was traced to the use of larger cans. Heat penetration was insufficient to kill the bacteria in the middle. Scale-up is a modification[25].

    Section 2.3, Incident 2 (page 16) describes a seemingly trivial change in design which had serious results.

    While many procedures are neglected or ignored as soon as managers lose interest, a few are followed religiously. For example, many studies have shown that spark-resistant tools (originally called non-sparking tools) have very limited value, and some disadvantages (see Appendix 7.3, page 124), but nevertheless many companies continue to use them for all maintenance work on plants handling flammable hydrocarbons.

Another example: many years ago I was responsible for a plant that manufactured acetone by the dehydrogenation of isopropanol in the vapour phase over a catalyst. As the catalyst aged, the temperature of the vapour leaving the reactor had to be raised but this reduced the catalyst life. I wrote an instruction stating that this temperature was not to be raised above a certain value (let us say it was 160°C) unless a catalyst change was imminent.

After I had left the plant, one of the reactors was adapted to make a different product from a different raw material using a different catalyst. My successor rang me up to say that the operators were very reluctant to raise the exit temperature above 160°C and did I know why?

Why are some procedures followed so scrupulously while others are abandoned at the drop of a hat? It seems that those that are easy to carry out are followed, while those that are not are allowed to lapse. We cannot make all procedures easy to follow, but if the safe way of carrying out a job is more difficult than the unsafe way, we should be on the alert. If we turn our backs for a moment the unsafe way will creep in.

THE NEED FOR PATIENCE

This chapter has stressed the need for greater management action. At the same time we should realise that everything cannot be done at once. Suppose a new manager (or safety adviser) sees equipment or a practice that is obviously unsafe. The equipment will take time to put right — modifications have to be designed and made — but why not change the practice right away? If we do not, someone may be injured. Unfortunately rapid change is usually impracticable, unless the procedure is very simple. If we find, for example, that the systems for preparing equipment for maintenance or controlling modifications are poor it is no use issuing an edict for immediate change. We have to persuade people that change is necessary, involve them in drawing up the new procedures, explain them and the reasons for them and discuss objections. All this takes time. Most people do not succumb on the spot; they have to be seduced by degrees. We cannot rule out this possibility that an accident may occur during the period of change but the fact that change is in the air may make people more aware of the hazards.

However, after a serious accident people do accept change by decree, as we saw in Section 2.1 (page 4).

7.9 THE OBVIOUS

The subjects discussed in this section are obvious only in retrospect. We are often so busy finding complex solutions to complex problems that we fail to realise that:

- What you don't have, can't leak.
- People who are not there cannot be killed.
- The more complicated a plant becomes, the more opportunities there are for human error and equipment failure.

WHAT YOU DON'T HAVE, CAN'T LEAK

The best way of preventing a leak of hazardous material is to use so little that it does not matter if it all leaks out, or to use a safer material instead. We cannot always find ways of doing this but once we start looking for them we find a surprisingly large number[9].

Until the explosion at Flixborough, UK, in 1974, which killed 28 employees[22,23] the attitude in the process industries was, 'There is no need to worry about large stocks of hazardous materials as we know how to keep them under control'. Flixborough weakened that confidence and ten years later the toxic gas release at Bhopal, India, which killed over 2000 members of the public[22], almost destroyed it. Since then companies, to varying extents and with varying degrees of success, have tried to reduce their stocks of hazardous materials in storage and process.

To do so requires a major change in the design process: much more consideration of alternatives in the early stages of design, as discussed in Section 6.5 on page 83, and this will not occur without encouragement and support from the senior levels of management.

Other industries have different problems but the principle is the same: can we avoid hazards, instead of trying to control them by adding onto our plants a lot of expensive protective equipment whch may fail or be neglected? Note that the expense is not just the initial cost of the protective equipment. The cost of testing and maintaining it, and the cost of the management effort needed to make sure that this takes place, can easily equal or exceed the initial cost. Trips and other complex protective equipment cost at least twice what you think they cost.

PEOPLE WHO ARE NOT THERE CANNOT BE KILLED

Just as material which is not there cannot leak, people who are not there cannot be killed. Those killed at Bhopal were living in a shanty town which had grown up close to the plant. Earlier in the same year 542 people were killed in a BLEVE (see Section 6.3 on page 76) in Mexico City. Most were living in a shanty town near the plant[26].

It is difficult in some countries to control development. Nevertheless plants handling large amounts of hazardous materials should not be built unless concentrations of people can be kept well away.

Fortunately disasters such as Bhopal can occur in only a few plants and industries but the principle of segregation applies everywhere. By sensible lay-out — for example, by keeping people and traffic apart or by not putting a workshop near a unit that might explode — we can reduce the chance that someone will be killed or injured.

THE MORE COMPLICATED A PLANT BECOMES, THE MORE OPPORTUNITIES THERE ARE FOR HUMAN ERROR AND EQUIPMENT FAILURE

The usual response to this statement is that complication is inevitable today. Sometimes it may be, but not always; Reference 9 describes many ways in which plants have been made simpler, and thus cheaper and safer. As with the reduction of stocks, the constraints are often procedural rather than technical. We cannot simplify a design if we wait until it is far advanced; we have to consider alternatives in a structured and systematic way during the early stages of design (see Section 6.5, page 83).

7.10   THE SPECIAL RESPONSIBILITIES OF THOSE AT THE TOP

Today everyone agrees that the safety record of a company, like its output, sales, quality and profit, depends on the ability of its directors and senior managers. By and large they accept this. They emphasise the importance of safety; they exhort their staff to do better and on the whole they do not stint resources. But often they do not give to safety the same detailed attention as they give to other management functions.

If output, sales, costs or quality are not up to the standard expected then directors and senior managers identify the problems and agree actions. They make regular checks to see that these actions are carried out and are having the desired effect. This is normal good management but is often not followed where safety is concerned. Instead exhortation replaces analysis, superficial views on accident causation are accepted without criticism and the lost-time accident rate is the only parameter seen.

When accidents have been investigated by a public inquiry, the reports do criticise the senior managers of the organisations involved. See, for example, the reports on the *Herald of Free Enterprise*[27], King's Cross[28], Clapham Junc-

115

LESSONS FROM DISASTER

tion[29] and Piper Alpha[30] as well as older reports such as those on Flixborough[23] and Aberfan[31]. (These reports, except Piper Alpha, are summarised in References 12 and 14 and, except the first, more briefly in Sections 4.1 and 4.5, pages 41 and 56 respectively.) Lawyers are involved in public inquiries and they seem to be skilled at uncovering the underlying causes of accidents. In comparison, most company reports, and many of those by regulatory bodies, are superficial and deal only with the immediate technical causes. The authors of company reports are, of course, reluctant to criticise their senior managers (it is easier to criticise those at the bottom of the pile) but a more important reason is a genuine failure to see that there is any action that the senior managers could take.

(In another respect official inquiries are less satisfactory: they often produce a long list of recommendations without any indication of their relative contributions to risk reduction or any cost-benefit analysis. See Section 7.7, page 107.)

What, then, are the safety problems that senior managers might usefully follow up or the attitudes they might inculcate? Obviously needs differ from company to company and many of the most important have been discussed already. They include:

- A new attitude towards human error (see Section 7.7, page 108).
- Reduction of stocks of hazardous materials (see Sections 6.5 and 7.9, pages 83 and 114 respectively).
- Simplification of plant design (see Section 7.9, page 114).
- More thorough investigation of accidents and dangerous occurrences (see Section 7.7, page 106).
- Better procedures for preparing equipment for maintenance and controlling modifications (see Sections 6.1 and 7.8, pages 70 and 109 respectively).
- Better communications: some sites may have problems that others have solved.
- Professionalism: reluctance to bring in experts or talk to those who know[32].

And, of course, learning and remembering the lessons of the past.

When discussing human errors in Section 7.7 (page 106), I said that we should not blame people who make errors but look for the reasons and, if necessary, provide better training. The same applies to senior managers. If they do not do as much as they might to prevent accidents, the reason is not that they do not care or cannot spare the time or money, but because they do not see that they could do more. Like many operators, they need training. But while there are many courses and packages for the training of operators and junior managers, the training of senior managers and directors is a neglected field. And it is neglected because the customers do not recognise the need for the service.

A report prepared for the Workers Educational Association in 1991[33] said that most accidents could be prevented by better management. It then concluded that if managers could prevent accidents and do not do so, they must be wicked and should be punished. It 'demanded' more prosecutions, harsher penalties and prosecutions for manslaughter rather than for offences such as 'failing to provide a safe plant'. It apparently did not occur to the author of the report that the failings of managers — like those of other people — might be due to incompetence rather than wickedness, and that managers are not always as well-trained as they might be. Managers are not supermen; they are like the rest of us.

Friedman writes, 'Those who have never really wielded power always have illusions about how much those who have power can really do'[34].

The Workers Educational Association usually supports the view that we should look for the underlying causes of law-breaking instead of putting law breakers in prison. If this applies to inner city rioters, why not to directors?

Of course, there is a place for prosecution and there are some people who will respond to little else. But the Workers Educational Association's 'demands' would prevent only a few accidents because they do not tackle the real problems.

Appendix 7.2 (see page 122) describes a prosecution which will have done nothing to prevent future accidents and merely distracted attention from the underlying causes.

REFERENCES IN CHAPTER 7
1. Fennell, D., 1988, *Investigation into the King's Cross Underground Fire* (HMSO, London, UK).
2. Marshall, V.C., 1992, *Applied Energy*, 42: 63.
3. Bond, J., April 1988, *Loss Prevention Bulletin*, No 080, 23. See also *Safety and Loss Control Management and the International Safety Rating System* (Det Norske Veritas, Norway), undated.
4. Allinson, J.T. and Faultless, C.G., Do you know where you are exposed?, *Management of Safety (Manosaf '91) Conference*, Institution of Chemical Engineers London & South East Branch, London, UK, 10–11 April 1991.
5. Literature available from the Liquefied Petroleum Gas Safety Association of South Africa, PO Box 456, Pinegowrie 2123, South Africa.
6. Davis, K.C., 1990, *Don't Know Much about History* (Avon Books, New York, USA), 237.
7. Health and Safety Executive, 1990, *The Peterborough Explosion* (HMSO, London, UK).

8. Kletz, T.A., 1992, *Hazop and Hazan — Identifying and Assessing Process Industry Hazards*, 3rd edition (Institution of Chemical Engineers, Rugby, UK; published in the USA by Taylor & Francis, Bristol, PA).
9. Kletz, T.A., 1991, *Plant Design for Safety — A User-Friendly Approach* (Hemisphere, New York, USA).
10. Health and Safety Executive, 1989, *Risk Criteria for Land-Use Planning in the Vicinity of Major Industrial Hazards* (HMSO, London, UK).
11. Kletz, T.A., 1992, in *Engineering Safety*, edited by D.I. Blockley (McGraw-Hill, London, UK), Chapter 15.
12. Kletz, T.A., 1988, *Learning from Accidents in Industry* (Butterworths, London, UK).
13. W.H. Doyle was probably paraphrasing H.L. Mencken, 'There usually is an answer to any problem: simple, clear and wrong' (quoted in Leo Rosten's *Book of Laughs*, 1986 (Elm Tree Books, London, UK), 363.
14. Kletz, T.A., 1991, *An Engineer's View of Human Error*, 2nd edition, (Institution of Chemical Engineers, Rugby, UK).
15. Kletz, T.A., 1988, *What Went Wrong? — Case Histories of Process Plant Disasters*, 2nd edition (Gulf, Houston, Texas, USA), Chapter 1.
16. Kletz, T.A., November 1976, *Chemical Engineering Progress*, 72 (11): 48.
17. Sanders, R.E., 1993, *Management of Change in Chemical Plants: Learning from Case Histories* (Butterworths, London, UK).
18. Pinney, S.G., June 1990, *Materials Protection*, 37.
19. *The Bulletin, The Journal of the Association of Petroleum Acts Administration*, July 1982, 20 (3): 35.
20. Rolt, L.T.C., 1974, *Victorian Engineering* (Penguin Books, London, UK), 90.
21. Health and Safety Executive, 1976, *The Explosion at Appleby-Frodingham Steelworks, Scunthorpe, 4 November 1975* (HMSO, London, UK).
22. Kletz, T.A., 1988, *Learning from Accidents in Industry* (Butterworths, London, UK), Chapters 8 and 10.
23. Parker R.J. (Chairman), 1975, *The Flixborough Disaster: Report of the Court of Inquiry* (HMSO, London, UK).
24. Capraro, M.A. and Strickland, J.H., October 1989, *Plant/Operations Progress*, 8 (4): 189.
25. Fore, H., 1991, Contributions of chemistry to food consumption, *Proceedings of 150th Anniversary Congress* (Royal Society of Chemistry, London, UK).
26. Kletz, T.A., 1988, *What Went Wrong? — Case Histories of Process Plant Disasters*, 2nd edition (Gulf, Houston, Texas, USA), Section 8.1.4.
27. Dept of Transport, 1987, *MV Herald of Free Enterprise: Report of Court No 8074: Formal Investigation* (HMSO, London, UK).
28. Fennell, D. (Chairman), 1988, *Investigation into the King's Cross Underground Fire* (HMSO, London, UK).
29. Hidden, A. (Chairman), 1989, *Investigation into the Clapham Junction Railway Accident* (HMSO, London, UK).

30. Cullen, W.D., 1990, *The Public Inquiry into the Piper Alpha Disaster* (HMSO, London, UK).
31. *Report of the Tribunal Appointed to Inquire into the Disaster at Aberfan on October 21st, 1966*, 1967 (HMSO, London, UK). For a shorter account see Kletz, T.A., 1988, *Learning from Accidents in Industry* (Butterworths, London, UK), Chapter 13.
32. Kletz, T.A., 1988, *Learning from Accidents in Industry* (Butterworths, London, UK), Chapters 4 and 5.
33. Bergman, D., 1991, *Deaths at Work — Accidents or Corporate Crime* (Workers Educational Association, London, UK).
34. Friedman, T.L., 1990, *From Beirut to Jerusalem* (Doubleday, New York, USA), 207.
35. Amundsen, R., quoted by Griffiths, T., 1986, *Judgement over the Dead*, (Verso, London, UK), 129.
36. Harvey-Jones, J., 1992, *Getting It Together* (Mandarin, London, UK), 162.

APPENDIX 7.1 — A FUTURE ACCIDENT REPORT

As a change from all the old accident reports quoted so far, here is an extract from a future official report:

'We acquit the directors and senior executives of the Company of any lack of commitment to safety. Their policy statement is unequivocal in the importance and priority it attaches to the subject. The Annual Report and speeches by directors, reported at length in the company newspaper, are equally clear and the chief safety officer has agreed that any resources he asked for have always been made available. However, we find that the directors have failed lamentably to achieve in practice what they proclaimed in principle.

'All the technical recommendations we have made in earlier chapters have been made before. Admittedly they are not collected together in one place but are scattered amongst many books, journals and old accident reports. But all these documents are available in the Company's libraries and we find it surprising that there was no system for collecting together those published and unpublished recommendations that are relevant to the Company's operations, drawing them to the attention of the staff, incorporating them in the Company's design standards and operating procedures and checking to see that they were being followed. Though the Company has been carrying out safety audits for many years, no attention has been paid to this point.

'Not only has the Company failed to learn from incidents elsewhere, but it has also failed to learn from those which have occurred on its own premises. They have, until now, caused few injuries but that has been due more to good luck than good management. Formal incident investigations have been held for many years and the neatly-typed reports look impressive, but we found when we read them that they were superficial. They dealt only with the immediate technical causes of the incidents but not with ways of avoiding the hazards or with the underlying weaknesses in the design process or management system; and they were too ready to lay the blame on human error without saying how such errors can be avoided in the future. This is not the fault of the investigating teams as they were too close to the detail to be able to see the wider picture. The responsibility lies with those senior managers who should have vetted the reports (or asked other outsiders to do so) but who seem to have given them only the most cursory examination.

'We have drawn attention in an earlier chapter to the fact that the disaster would not have occurred, or would have been much reduced in scale, if the amount of hazardous material in progress had been lower. The technology for reducing inventories was available but no-one was encouraged to pursue it actively, the feeling being that hazardous materials could be kept under control. We are surprised that the Board took so little interest in the most obvious way of reducing hazards and did not ask for regular returns of inventories and the progress made in reducing them. The only safety figures reported on a regular basis were the lost-time accident rates, though middle and junior managers have told us that they considered them of little value.

'We find that the underlying cause of the disaster was the failure of the directors and senior executives to carry out a task with which they had been entrusted and which they accepted as part of their responsibilities. This was not due to any lack of commitment but to a failure to appreciate the way they should have carried out the task. The Company's training programmes are, in many ways, impressive but this Company, and perhaps the whole industry, has no programme for training directors. We were shown scores of safety training programmes, both in-house and external, but not one was intended for senior managers. One training consultant told us that he never arranged courses for them as they were unlikely to attend.'

APPENDIX 7.2 — TRAIN DRIVER ERROR

A railway accident shows very clearly how concentration on the immediate cause of an accident and blaming the person involved can distract attention from the underlying causes and the responsibilities of senior managers.

In March 1989 a British Rail train driver passed a caution signal without slowing down; he did not brake until he saw a red signal and by then it was too late and his train crashed into the rear of the train in front. Five people were killed and 88 injured. There have been many similar accidents and as the driver is at the front of the train he is more likely than anyone else to be killed. On this occasion he survived and in September 1990 he was prosecuted for manslaughter, convicted and sentenced to six months' imprisonment. In November 1990 the Court of Appeal confirmed the conviction but reduced the sentence to four months of which two were actually served.

The decision of the Courts surprised many students of human error and industrial safety and weakened their confidence in the understanding of the judiciary.

The driver's error was not due to lack of training or to a deliberate decision to pass the signal or drive carelessly. (As already stated, in accidents of this type the driver is the most exposed person in the train and is usually killed.) His error was due to one of those momentary slips or lapses or attention which afflict us all from time to time and which we cannot avoid. The responsibility for the accident lies with those who failed to provide the equipment which could have prevented it (they have recently agreed to do so), not with the man at the bottom of the pile who could not pass the blame onto someone below him.

In the present British Rail system of automatic train control, a hooter sounds in the cab when a train approaches a signal at red (stop), yellow (caution) or double yellow (advance caution) and if the driver does not cancel the hooter, by pressing a button, the brakes are applied automatically. It is, of course, possible for the driver to cancel the hooter and take no further action, and if the train is already going slowly this may be the right thing to do. On busy lines drivers are constantly passing yellow and double yellow signals and pressing the button can become almost automatic; without realising what they are doing drivers cancel hooters and carry on. This became increasingly common during the 1980s and the Railway Inspectorate urged British Rail to install more advanced equipment (called automatic train protection) which cannot be cancelled by the driver and which will automatically apply the brakes before the train passes a red signal. In 1988 British Rail agreed to install this equipment by 1998.

The managers who did not provide the equipment which could have prevented the accident were not, of course, unconcerned about safety. At worst they did not understand the nature of human error (railway managers have been known to say, 'Drivers are paid to obey signals[1]); at best they may have decided that more lives would be saved if the railway's resources were spent in other ways.

The official report on the accident[2] points out that the signal involved had been passed at red on several previous occasions and a 'repeater' has now been installed to improve its visibility.

The last time a driver was imprisoned for passing a signal at danger was, I think, in 1913 following an accident at Aisgill on the Settle-Carlisle line[3]. The driver was distracted by problems with his engine and the Inquiry criticised the management for the strictness with which they punished delay and incorrect treatment of engines. As a result of this policy drivers were more afraid of being late than of having an accident. The accident led to the increased use of automatic train control, already in use on the Great Western Railway.

REFERENCES IN APPENDIX 7.2
1. Hall, S., 1987, *Danger Signals* (Ian Allan, London, UK), 126.
2. *Report on the Collision that Occurred on 4th March 1989 at Purley in the Southern Region of British Railways*, 1990 (HMSO, London, UK).
3. Schneider, A. and Mase, A., 1970, *Railway Accidents of Great Britain and Europe* (David and Charles, Newton Abbot, UK), 49.

APPENDIX 7.3 — A NOTE ON SPARK-RESISTANT TOOLS

These tools, formerly called non-sparking tools, seem to be regarded as a sort of magic charm to ward off accidents. Though many companies use them, a series of reports, from 1930 onwards, have shown that their value is very limited.

Impact of steel hand tools on steel may ignite ethylene, hydrogen, acetylene and carbon disulphide but not most hydrocarbons. It may seem, therefore, that there is a case for using spark-resistant tools on plants which handle these substances but people should never be allowed to enter a flammable atmosphere, much less work in one, as it may ignite while they are present. The most that should be allowed is putting the hands, suitably protected, into a small area of flammable mixture (say, up to 0.5 m across) immediately surrounding a leaking joint, in order to tighten it. If the leaking gas is one of those mentioned then a spark-resistant hammer should be used. Spark-resistant spanners are said to be less effective than steel ones and there is no need to use them[1].

In exceptional circumstances someone might be allowed to enter a flammable cloud to close a valve and stop a leak which would otherwise spread. If possible, he should be protected by water spray. No-one should be asked to do so.

These comments apply only to mixtures of flammable vapour and air, not to solid explosives.

REFERENCE
1. Powell, F., 1986, *Reprint No 1110* from *Gas, Water, Wastewater*, No 6 (Swiss Gas & Water Industry Association, Zurich, Switzerland).

8. AN ANTHOLOGY OF ACCIDENTS

'In learning the sciences, examples are of more use than precepts.'
Isaac Newton[1]

'When we are children we learn history by quantities of little unrelated pictorial anecdotes that adhere with such obstinacy that no amount of scholarly research later on will ever quite dislodge them.'
G.B. Stern[2]

'Modern laws set out the rules: ancient laws set out examples ... one is expected to deduce the general principle which could then be applied to other cases ...'
M. Hilton and G. Marshall[3]

Advice and theories are hard to remember. In contrast, if a story catches our imagination it may stick in our memories. So in this chapter I describe some accidents which will, I hope, lodge in our memories and will (like Wordsworth's daffodils) 'flash upon that inward eye' if we find ourselves in a similar situation. If we cannot remember the advice we were given on the action to take before we modify the plant or engage a contractor, perhaps we will remember what happened to people who failed to follow the advice.

In other chapters (and in *Learning from Accidents in Industry*[4]) I have tried to show that we should look below the obvious technical causes of an accident for the underlying causes, for ways of avoiding the hazard and for weaknesses in the management system. In contrast, in this chapter, I emphasise a particular feature of each accident in the hope that, like Alfred and the cakes or Bruce and the spider, it will stick in the memory.

8.1 RELUCTANCE TO RECOGNISE THAT SOMETHING IS WRONG

The predominant feature of this incident was the reluctance of many of those involved to recognise, and then admit, that anything was wrong (see also Section 2.3, Incident 3 on page 17 and Section 7.4 on page 99). It is based on Joyce Egginton's book, *Bitter Harvest*[5].

Nutrimaster, a food additive for animals, and *Firemaster*, a fire-retardant chemical containing polybrominated biphenol, a toxic chemical, were normally packed in differently coloured bags but in 1973 the manufacturer was unable to obtain the usual bags and used similar brown paper bags for both. A manufacturer of animal feeding stuffs was suppled with *Firemaster* instead of *Nutrimaster* and incorporated it in his products. There was an epidemic of illness in the surrounding farms, in people and animals, and thousands of animals died or had to be destroyed.

It was nine months after the first animals became ill before the cause was discovered and it might have been longer if one of the farmers affected had not been trained as a chemical engineer.

Those involved included the feed supplier, several vets, the Michigan Department of Agriculture, several groups in the United States Federal Department of Agriculture and several universities but no-one was in overall charge, there was no coordination, no-one knew the whole story and for a long time each affected farmer did not know that others had sick animals. Each authority felt that it was not really their problem. The doses received were variable and the effects inconsistent and so everyone was ready to blame the farmers for mismanagement.

When the cause of the sickness was found the authorities were slow to recognise and admit the scale of the problem and senior people made reassuring noises.

Later it was found that low levels of contamination produced delayed effects, and again there was a long delay before the authorities would admit that this was a serious problem. Even when they admitted that low levels of contamination had produced effects in animals, they were reluctant to admit that this had also occurred in humans.

Some of the farmers felt that the authorities were deliberately refusing to recognise the obvious but, when something goes wrong, we all instinctively look for explanations that make someone else responsible. Egginton says (page 150) that the authorities did not lie but, 'They just did not tell the whole story — the automatic bureaucratic reaction to a situation which does not look good and cannot be remedied'. One farmer commented, 'It was not what was done that was wrong, but what was not done by a number of people in authority who did not realize the magnitude of the problem' (page 152).

Lack of clear responsibility has been a factor in many other major incidents. For example, Lord Cullen's official report on the fire and explosion on the Piper Alpha oil platform in the North Sea in 1988, in which 167 people were killed, said:

'No senior manager appeared to me to 'own' the problem [blockages in water deluge lines] and pursue it to an early and satisfactory conclusion.'[6]

Similarly, during the official inquiry into the loss of the cross-Channel ferry *Herald of Free Enterprise* in 1987, in which 186 people were killed, one director told the Court of Inquiry that he was responsible for safety while another director said that no-one was[7]. See also Section 8.4, page 128.

8.2 LABELLING SHOULD BE CLEAR AND UNAMBIGUOUS

The *Nutrimaster/Firemaster* incident shows the dire results that can follow if we do not give enough attention to prosaic matters such as making sure that labels are distinctive and unambiguous. Here is another incident.

A lady was admitted to hospital for an operation on her lower right leg. The house officer marked the limb with a proprietary skin marker. She then promptly crossed her legs, and printed the arrow on the other leg. Fortunately, before carrying out the operation, the surgeon saw what had happened[8].

8.3 REASSESS HAZARDS FROM TIME TO TIME

To what extent should companies re-examine processes which have been in use without problems for a number of years but which might, in the light of new knowledge or new techniques, present some hazards?

In June 1976 a devastating explosion occurred at Dow's King's Lynn factory in a vessel in which zoalene (3,5-dinitro-*ortho*-toluamide), a poultry food additive, had been left after drying. Afterwards tests by accelerating rate calorimetry (ARC) showed that zoalene could self-heat to destruction if held at 120–125°C for 24 hours.

Dow had only recently developed the ARC technique and had started to test the chemicals they handled. At the time of the explosion they had tested 5% of them. The Health and Safety Executive (HSE) agreed, in its report on the explosion[9], that there had been no reason to give zoalene priority as other tests and 17 years' manufacturing experience had given no inkling of its instability. The HSE concluded, 'There appears to be no substantial grounds for criticising the Dow management or operating personnel for undertaking and conducting the operation that led to the explosion in the way they did'. The conclusion would, of course, have been different if the zoalene had waited until today before blowing up and no tests had been carried out.

In the late 1960s ICI designed plants for the vapour phase oxidation of ethylene (to vinyl acetate and ethylene oxide). To meet ICI's safety standards the control systems had to be made more complex than any previously installed on a chemical plant; they were of almost nuclear complexity (see Section 6.4, page 81)[10]. Formaldehyde had been manufactured by the vapour phase oxidation

LESSONS FROM DISASTER

of methanol for many years without any problems. Should the process be looked at again in the light of the experience gained in designing the new oxidation plants? An investigation, undertaken with some trepidation, showed that the control system was, in fact, inadequate but this time the problem was solved by an ingenious redesign rather than by extra complication[11].

After a runaway reaction, during the 1970s, a director said that all his company's processes must be subjected to a hazard and operability study. His staff had to explain to him that this was quite impracticable on any realistic timescale, but they went on to develop a systematic technique for putting the processes into a sort of batting order. The most hazardous, and those about which least was known, could then be studied first. (This, incidentally, shows how an impracticable suggestion can have fruitful results if it stimulates other people into thinking of other but more practicable solutions. In a hazard and operability study we should not be reluctant to put forward any impracticable causes, consequences or solutions that occur to us; they may stimulate other members of the team who may produce more practicable variations. In science, false ideas can prove useful if disproof leads to greater knowledge.)

A similar method has been devised for the rapid ranking of hazards, to establish priorities for more thorough investigation[12,13]. In another UK company every plant is, at the time of writing, being reviewed by a team of three or four people on the lookout for sources of leakage (such as drain valves, vents and machines not fitted with emergency isolations), non-compliance with current standards and safety critical procedures. Each study takes a few days and usually finds some hazards that can be put right fairly quickly and easily and others that justify more detailed study by hazop and/or hazan. This company is probably the exception, however, and in the chemical industry as a whole I suspect that much more needs to be done. While much has been published on hazard reviews for new plants, much less is available on similar reviews for old plants.

8.4   WATCH CONTRACTORS CLOSELY

I have argued elsewhere[14] that the commonest cause of major leaks is pipe failures and that the commonest cause of pipe failures is the failure of construction teams to follow instructions or to do well, in accordance with good engineering practice, details that are left to their discretion. The most effective single action we can take to reduce pipe failures, and thus major leaks, is to specify designs in detail and then inspect thoroughly during and after construction — more thoroughly than has been customary in the past — to make sure that the design has been followed and that details not specified in the design have been carried out in accordance with good engineering practice.

The report on a flood in the Victoria and Albert Museum in London in March 1986[15] illustrates these points and also gives a better insight into the underlying causes than most reports. Incidents in other industries are often instructive because, not being involved, we see things more clearly.

A clamp joining a temporary water pipe to the building pipe came undone, flooding a basement and causing £250 000 damage to the contents. The contractor had used a type of clamp that was intended for repairing cracked or corroded pipes but not for joining two pipes. In addition, the clamp was designed for pipes of 115–124 mm diameter but the building pipe was 130 mm diameter. Gross carelessness on the part of the contractors? But read on. The clamp they used was immediately available. The type recommended by the manufacturers for joining two pipes was on four months delivery, even longer for the non-standard size required (130 mm pipe is no longer standard). What were the contractors expected to do? We are left wondering what is 'custom and practice' in the industry, and how many similar jobs had been bodged in the past, a point not discussed in the report. Many years ago, after an explosion on a chemical plant, a survey of flameproof electrical equipment showed that many bolts were missing. Carelessness on the part of the electricians? It was found that the special bolts and screwdrivers required were not in stock. Lost tools and bolts could not be replaced (see Section 4.3, page 49).

After assembling the pipe the contractors turned on the water and watched it for 1½ hours before going home. However, during the night the water pressure rose by a third and the clamp came adrift. Apparently no-one was aware that the pressure rose when demand fell, another example of an accident caused by ignorance of well-known knowledge — or failure to apply it.

A superintending officer (appointed by the government department responsible for the building, the Property Services Agency [PSA]), and a resident engineer were on site at the time the joint was made. According to the report they would not have been expected to examine individual joints in detail, and it is doubtful whether they would have spotted anything wrong if they had examined the joint.

Basements are always liable to be flooded so it does not seem very clever to store valuables there. ('What you don't have, can't be damaged.') However, the Museum staff may have had little choice. If basements must be used, the report recommends water level alarms and water-tight doors.

There seems to have been little systematic consideration of the hazards and one of the recommendations made in the report is that there should be a formal risk assessment during design. It also states that temporary work should be given better scrutiny during design and better supervision and testing during during construction.

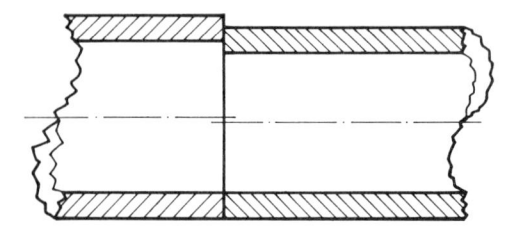

Figure 8.1 Two pipes of slightly different diameter were welded together.

Nobody can be blamed for the incident. Everybody did the job they were expected to do, with the constraints imposed by availability of materials, but it was nobody's job to ask if the existing set-up was adequate, to audit the management system, or to introduce a systematic procedure of the hazop type for identifying hazards and operating problems. It is not clear whether the Museum, the PSA or the consulting engineers were in overall control.

Better control of contractors could have prevented many other accidents. For example:

- A construction worker misread a drawing and cut a hole in a pipe at the wrong place. When he realised his error he welded a patch over the hole and said nothing. The line was radiographed but this weld was omitted as the radiographer did not know it was there; the line was then insulated, hiding the patch. The weld was sub-standard and leaked; several men were affected by phosgene and one was nearly killed.
- Two lengths of 8 inch (200 mm) pipe which had to be welded together did not have exactly the same diameter. The welder joined them so that there was a step between them along part of the circumference, as shown in Figure 8.1. The pipe was then insulated, again hiding the defect, and the poor workmanship was not discovered until the insulation was removed for inspection of the pipe ten years later. Failure of the pipe could have caused a Flixborough-like incident as it carried petrol at 150°C and 10 bar gauge.
- An old pipe was reused. It looked OK but it had operated for twelve years at 500°C and had used up much of its creep life. When it failed in service, a leak of high pressure gas produced a flame 30 m long.

At first sight these incidents, and I could quote many more, support the view held by many operating staff that construction contractors are incompetent and irresponsible. But did the men involved in these incidents know what substances would go inside the pipes they were constructing? Did they know what could happen to other people if their workmanship was below standard? That other people's lives were in their hands? I have suggested to construction engineers that they should tell their workforce about the substances that will be used in the plants they build and what will happen if the plants leak, but most of them dismiss the suggestion as impracticable. 'You should see the type of labour we employ', they say, 'here today and gone tomorrow. All they are interested in is a quick buck'. Are they right?

Finally, an example from history to show what can happen if contractors are treated as second class citizens. The 12th century Crusader castle of Belvoir in Israel, overlooking the Jordan valley, seems impregnable. There are cliffs on three sides, a deep dry moat and several defensive walls. How, the visitor wonders, did such a castle ever fall to the enemy?

Many of the defenders were mercenaries, that is, contractors. Their loyalty was suspect, so the innermost fortification was reserved for the Crusader knights. The mercenaries knew that they were ultimately expendable and when the going got rough they were tempted to change sides. The knights knew they might do so and it affected their own morale. They sued for terms[16].

Sometimes contractors are not supervised adequately because those who are supposed to supervise them have too many other things to do. The next incident illustrates this point.

8.5   DO NOT OVERLOAD SUPERVISORS

Many people believe that they work best when they are moderately overloaded and many managers deliberately set out to give their staff rather more work than can conveniently be done in a normal working week. This encourages people to make the best use of their time and can teach them to distinguish essential jobs from inessential ones. However, it is too risky a policy to apply to those in immediate control of hazardous plants.

Two jobs had to be carried out at the same time in the same pipe-trench. The first was construction of a new pipeline. The process foreman issued a permit-to-work, including a welding permit, at 08.00 hours, valid for the whole day. At 12.00 hours a permit was requested to remove a slip-plate from an oil line. The line had been drained but a little oil remained in it, as the drain point was a few inches higher than the bottom of the joint to be broken — not a good feature. The foreman was busy on the operating plant, 500 m from the pipe-trench, at the time and authorised the removal of the slip-plate without visiting the site. He judged that the contractors would by now be a safe distance from the site of the slip-plate. (The usual safe distance required between a source of

LESSONS FROM DISASTER

ignition and a flammable liquid is 15 m [50 feet] and this includes a safety margin; the contractors and the slip-plate were in fact 20 m [65 feet] apart.)

However, had the foreman visited the pipe-trench he might have noticed that it was covered by pools of water. The few gallons of oil that came out when the slip-plate was removed spread over the surface of the water for a far greater distance than 15 m and were ignited by the welder working on the pipeline. The man who was removing the slip-plate was killed.

The actual distance between the two jobs (20 m) was rather close to the minimum normally required (15 m) and the foreman was fortunate that his judgement of the distance, made without visiting the site, was correct. He did not visit the site because he felt that his main job was to supervise the operation of the plant and current problems required his presence in the control room. His judgement may have been wrong but he did not make his decision in a vacuum; he would have been affected by the company culture and perhaps he felt that he would be criticised if he had left the plant or held up either of the jobs in the pipe-trench (compare Section 7.1, last two paragraphs, page 94).

After the accident a special day foreman was appointed to supervise the construction contractor. It was realised that the operating foreman could not give both the plant and the contractors the attention they needed.

Of course, had the foreman visited the site he might not have realised that oil spilt on water will spread a long way, though there have been some spectacular incidents. In 1970 35 tonnes of petrol were spilt onto the Manchester Ship Canal; 2½ hours later and 1 km away the petrol caught fire and six men in a small ferry boat were killed[17]. About 1942 some petrol was spilt in Lagos Harbour, Nigeria; 1½ km away a sailor on a corvette lit a cigarette and threw the match into the sea. It caught fire and primed depth charges attached to the corvette and another one close by exploded. Forty people were killed and damage was extensive[18].

8.6 DO NOT CHANGE A DESIGN WITHOUT CONSULTING THE DESIGNER

In the Hyatt Regency Hotel in Kansas City, as in many Hyatt hotels, the lobby was the full height of the building and was criss-crossed by several horizontal walkways. One evening in 1981 the walkways were crowded with people watching the dancing on the floor of the lobby below when two of the walkways collapsed; 114 people were killed and nearly 200 injured. The reason for the collapse was a small change in a design detail[19].

The walkways were supported by metal rods attached to the ceiling. According to the design the rods passed through holes in the upper walkways,

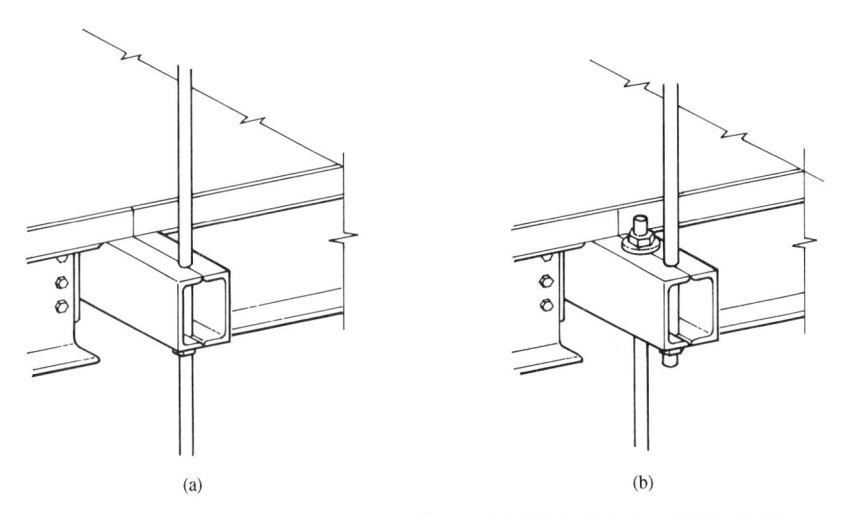

Figure 8.2 Method used to support a walkway. (a) Original design. (b) As built.

as shown in Figure 8.2(a), and the walkways were attached to the rods by nuts and washers. It is not clear how the nuts were to be fitted to the rods unless the whole of the distance between the nuts and the ends of the rods was threaded. Perhaps for this reason, perhaps to make assembly easier, the design was changed to that shown in Figure 8.2(b). This imposed a greater strain on the nuts and washers, a strain that they were unable to bear.

An analogy may make the reason for the collapse easier to follow. Suppose two people are hanging onto a vertical rope. If the rope is strong enough and their grips are tight enough they will not fall. Now suppose that instead of one long rope we have two short ropes. One person grips the upper rope and the lower rope is tied round his feet and gripped by the second person. The strain on the upper person's grip is doubled and may be too much for him.

The people who made the small change in design did not realise that they were increasing the load on the upper nuts and washers.

Section 7.8 on page 111 describes some other modifications that had serious unforeseen results.

8.7 WHAT SHOULD WE DO WITH SAFETY EQUIPMENT THAT STOPS US RUNNING THE PLANT?

Safety equipment sometimes operates when it should not, and upsets or prevents plant operation. For example, a trip operates although conditions are normal and shuts down the plant. In order to maintain production the operators disarm the

LESSONS FROM DISASTER

protective equipment and manage without it. The managers are not told and perhaps they prefer not to notice, as redesigning the protective equipment can be troublesome and expensive. In this situation it may be useful to recall the following story.

Everyone knows the San Francisco cable tramway. There is a similar one up the Great Orme in Llandudno, North Wales, but unlike the San Francisco one the cars are permanently attached to the cables. When it was built in 1902 it was fitted with an emergency brake held off by the tension in the cable. If the cable broke, the tension was released and powerful springs gripped the metal cover of the cable groove. They held so tightly that it took several hours to free them.

Clumsiness with the controls could cause the cable to jerk; the emergency brakes operated and the tramway was out of action for several hours. As a result, in 1905, without the knowledge of some of the directors, the emergency brakes were removed.

For 27 years nothing went wrong. Then in 1932 the steel drawbar on one of the tramcars suddenly broke. The car became detached from the cable and the driver could not stop it with the normal brakes. It ran down the hill, left the rails and ran into a wall, killing the driver and a 12-year-old girl. Ten passengers were injured.

The tramway was shut down by a government inspector for nearly two years while a new emergency brake was designed, tested and installed. The insurance company refused to pay any compensation to the injured passengers, on behalf of the company, as the safety equipment had been removed. The company was not earning any money and went bankrupt.

The official enquiry brought to light the fact that a similar drawbar had failed ten days earlier. But on this occasion the car was stopped with the normal brakes and the incident was ignored. Another familiar story. The two drawbar failures were due to the fact that the manufacturers had been given a wrong specification for the steel and had not been told the purpose for which they were to be used[20].

8.8 NOT ALL TRIPS ARE SPURIOUS

As we have seen, safety equipment sometimes operates when it should not. After this has happened a few times everyone assumes that that the next trip or alarm is false. Unfortunately it may be genuine. Next time you are tempted to disbelieve what the safety equipment tells you, remember the following stories.

The first incident occurred in a hospital in Zaragoza, Spain. An electron beam accelerator, used to irradiate cancer patients, broke down. After repair the energy level indicator showed 36 MeV when the energy level selection keys were set for lower levels. The operators assumed that the needle was stuck at 36 MeV and carried on.

The needle was not stuck. The machine delivered 36 MeV whatever level was selected and some patients got 3–7 times more radiation than their doctors had prescribed. The beam was narrower than it should have been and the radiation went deeper.

What went wrong?

- The repairs had been bodged, though it is not clear whether the contract repairman did not know what to do or simply carried out a 'quick fix'.
- The hospital physics service was supposed to check, after repairs, that the energy level selected and that indicated were in agreement. They did not check as no-one had told them there had been a repair.
- The physics service was also supposed to carry out routine checks every day but as few, if any, faults were found the test interval was increased to a month. I doubt if anyone calculated the fractional dead time or hazard rate; the report does not say.
- A discrepancy between the energy level selected and that indicated should trip the machine. However, it was easy to by-pass the interlock by changing from automatic to manual control[21].

The incident was not simply the result of errors by the operating, repair or physics staff. They had been doing the wrong things for some time but no-one had noticed (or if they had noticed, they did nothing). To quote from the official report on the sinking of the cross-Channel ferry *Herald of Free Enterprise* at Zeebrugge in 1987, 'From top to bottom the body corporate was infected with the disease of sloppiness'[7.]

The second incident occurred at Alton Towers leisure park in Staffordshire, UK. A car on a pleasure ride developed a fault and the system was shut down automatically. The operator could not see the stranded car, assumed the trip was spurious and restarted the ride with an override key. There were several collisions and six people were injured. The company had to pay fines and costs of almost £3500. The operator's training was described as 'woefully inadequate'[22].

If operators regularly disbelieve safety equipment, perhaps it should be kept in better order. We cannot expect people to believe alarms which are wrong nine times out of ten.

8.9   'IT MUST BE SAFE AS I'M FOLLOWING ALL THE CODES'

Unfortunately not, as the following story shows. A hydraulic crane collapsed onto a petrochemical plant, fortunately without causing any serious damage. The collapse occurred when the driver tried to lift too much weight for the particular length and inclination of the jib, but the alarm bell did not sound. The crane was fitted with all the alarms required by the then current code of practice, and they were all in working order.

A mechanical strut crane has three degrees of freedom. The angle of the jib can be altered, the jib can be rotated and the load can be raised and lowered. An alarm sounds if the driver tries to lift too big a weight for the particular jib angle.

A hydraulic crane has an extra degree of freedom. The length of the jib can be altered.

In the incident just described the driver set the angle of the jib and then extended the jib to its maximum length. The crane was then so unstable that it tipped up as soon as a small weight was lifted. Because the hydraulic crane has an extra degree of freedom, it needs an extra alarm. But it was not fitted with one — only with the minimum number required by the code of practice which had been written before hydraulic cranes came into general use.

So do not relax because you are following the rules or codes; they may be out of date.

Another example: in 1987 the cross-Channel roll-on/roll-off ferry *Herald of Free Enterprise* sank because it left harbour with the bow doors open. There was no indicator or alarm to tell the captain that they were open and no-one had to report that they had been closed[7]. The ferry company originally used ferries with visor doors, which lift up like a knight's visor, and can be seen from the bridge. When they changed to clam doors, which could not be seen from the bridge, no-one asked if a change in equipment or procedure was necessary[23].

8.10   LITTLE-USED EQUIPMENT BECOMES FORGOTTEN

In the incident decribed in Section 2.2, Incident 1 (page 10), four men were badly burned while trying to isolate a leak of propylene. Afterwards the equipment was resited and fitted with remotely-operated emergency isolation and blow-down valves so that any future leaks could be stopped from a safe distance.

Eight years passed before another leak occurred. It produced a visible cloud of propylene vapour about 1 m deep. (The visible part of the cloud was actually condensed water vapour.) Two experienced foremen completely forgot about the emergency equipment and entered the cloud to isolate the leak. Fortunately it did not ignite. Afterwards, when one of the foremen got back to his office, he realised the risk he had taken and complained that he should not be expected to take such risks. In his eagerness to maintain production he had completely forgotten that equipment had been provided to avoid the need for such risk-taking.

On another plant the cooling water system was supplied with town water which was added to the cold well from which the cooling water circulation pumps took suction. There was also a supply of less pure water for use in emergencies. One day the supply of town water failed, the level in the cold well fell and the foreman carried out a crash shutdown of the plant. He had forgotten about the emergency supply which had not been used for many years. The valve was operated every week, so that it did not become stiff, but by one of the operators, and the foreman was not involved.

A radio broadcast[24] described a similar incident on the railways. In 1898 an engine coupling rod broke and punctured the boiler. Disregarding their own safety, the driver and fireman remained on the footplate until they had applied the brakes and prevented the train running away. They were badly scalded and died from their injuries. Their heroism was widely praised. However, the train was fitted with vacuum brakes; without a steam supply the vacuum would leak away and the brakes would be applied automatically. Vacuum brakes had only recently been made compulsory (following the death of 80 people, including many children, in a runaway train accident in Armagh, Ireland in 1889[25]) and the driver and fireman probably forgot that they were fitted.

8.11   CROWD CONTROL

Failures of crowd control have caused several incidents in recent years, notably in 1989 when too many people were allowed into a football ground in Hillsborough, Sheffield, and 99 of those at the front were crushed to death. Such accidents are nothing new. A worse one occurred in Sunderland in 1883. Two thousand children, aged four to fourteen, attended a theatrical entertainment in the Victoria Hall (admission 1 penny), half of them downstairs and half in the gallery. When the show was over the performers threw presents into the audience. The children upstairs realised they were going to miss out and started to rush downstairs. There was a pair of swing doors at the bottom of the stairs but not ordinary swing doors; after they had opened about 0.5 m (20 inches) bolts dropped into slots in the floor and the doors automatically locked themselves. They had been designed this way so that patrons could be let in one at a time under control and no-one could get in free.

The leading children were jammed against the doors by those behind them. Some fell and others fell on top of them. Altogether 190 died in the

LESSONS FROM DISASTER

mountain of struggling children and many more were rescued unconscious. The *Daily Telegraph* was unwilling to apportion blame but other newspapers had no hesitation and spoke of 'the deplorable blunder' of packing so many unsupervised children into the gallery and the 'reprehensible, almost iniquitous, folly' of installing such doors. According to one report the doors had been deliberately locked in the half-open position by a man with a box of presents who wanted to hand them out one at a time[26].

This incident ought to be known to everyone responsible for buildings in which large numbers of people, especially children, are gathered. I wonder how many of them have heard of it.

Another such incident occurred in the Church of the Holy Sepulchre in Jerusalem on Good Friday 1834. The church was packed with pilgrims attending the ceremony of the holy fire, in which candles inside the sepulchre itself are miraculously lit, or so the pilgrims believed. It was described by the British Consul, Robert Curzon, who was among those present.

Many people collapsed, from fatigue and asphyxiation, as they had been standing all night in great heat and numerous candles had been lit, further reducing the oxygen content of the air. Some of those who collapsed were trampled underfoot. Other people panicked and tried to fight their way out; the guards used their butts and bayonets freely and got the best of the battle. Altogether 500 people died. Robert Curzon got involved in the fighting but had the sense to fight his way away from the door, back to the inside[27].

8.12   DIFFERENT PEOPLE — DIFFERENT VIEWS

However impartial we try to be — and we do not always try — our views are affected by our background and responsibilities; our first reaction, when there has been an accident, is to think of all the things that other people might have done to prevent it taking place. The following account shows how different people might react to the same accident.

The liquid in a pair of pumps (one working, one spare) decomposed on standing with the production of gas. To prevent damage to the seal the plant instructions said that the pumps should be isolated and drained when they were not in use, and the drain valve left open in case the isolation valves were leaking. Before starting up a pump, operators should close the drain valve. This was pointed out to new operators during their training.

One day the operator on duty — a young, enthusiastic man who had completed his training only a few months earlier — started up a pump without closing the drain valve. He was hit on the leg by liquid that came out of the drain valve and splashed off the baseplate. The liquid was corrosive and burnt his leg;

he was absent from work for two days. Only the plant manager's comments have been preserved but other people might well have commented as follows:

FOREMAN
'The accident was entirely due to the carelessness of the injured man who failed to follow instructions. He has been told to do so in future. No further action is necessary.'

PLANT MANAGER (SUPERVISOR IN THE USA)
' ... it was carelessness which led him to open liquors to the pump with the vent open ... Labels will be provided to remind operators that certain pumps may be left with open vents. Process supervisors will be told to ensure that all operators, and particularly those with limited experience, are fully aware of the elementary precautions they are expected to take in carrying out their duties.'

INJURED MAN
'I forgot, but it's easy to forget that there is something special about a particular pump when you are busy and trying to do everything at once. If the manager spent more time on the plant (including a few night shifts) and less time in his office, he might understand this.'

SHOP STEWARD
'Blame the working man, as usual. There should be an interlock to prevent the isolation valves being opened before the drain is closed.'

SAFETY OFFICER
'It is inevitable that sooner or later someone will forget that there is something special about certain pumps and start them up in the usual way. The labels will help but will soon become part of the scene and may not be noticed. Interlocks would be expensive. Can a relief valve be fitted to the pump and can the drain be moved so that it will not splash the operator?'

DESIGN ENGINEER
'No-one told me that the drain valve on this pump had to be left open. If I had known I might have arranged it differently. According to our Engineering Standards relief valves are fitted only on positive pumps, not on centrifugal ones.'

The plant manager's comments are taken from his typed report and suggest that he was rather reluctant to make any changes. The accident occurred some years ago and today most managers would, I think, be less willing to quote 'carelessness' as the cause of an accident and would be more willing to change the design. I suspect, too, that at the time many safety officers would have agreed with the manager.

8.13 WE CAN LEARN FROM A REPORT EVEN IF ITS CONCLUSION IS WRONG

When reading accident reports we should distinguish between the facts — what happened — and the cause, which usually involves some degree of speculation. We may be convinced that the cause reported is unlikely, even impossible, but there may still be lessons to be learned from the facts. The following light-hearted story illustrates this theme.

According to ancient reports, Daedelus designed and constructed wings so that he and his son, Icarus, could fly out of the labyrinth in which they were trapped. The wings were made from feathers fixed to a frame with wax. According to Ovid, Daedelus warned Icarus that the equipment, like all equipment, had limitations:

"'My Icarus!" he says, "I warn thee fly
Along the middle track: nor low, nor high;
If low, thy plumes may flag with ocean's spray;
If high, the sun may dart his fiery ray.'"

Icarus flew too high, the heat of the sun melted the wax and he fell into the sea — known to this day as the Sea of Icarus — and was drowned (Figure 8.3).

We now know that as one goes higher it gets colder, not hotter, and so the official explanation cannot be the correct one. Stephen Barclay has suggested that the wax did not melt but failed as a result of brittle fracture[28]. Wax was a good material of construction for securing the feathers, as it yields slightly under increased strain, but if it falls below its ductile-brittle transition temperature, strain produces cracks which spread — slowly at first, and then rapidly.

Many years later, in 1953/4, similar fatigue cracks caused several failures of another flying machine — the Comet, the first commercial jet aircraft. In this case, fatigue was not the result of low temperature but of stress concentration at the corner of a cabin window. Low temperature has, however, caused fatigue failures of ships and process equipment. The first all-welded ships, with no rivetted joints to interrupt crack-propagation, were OK in warm waters but failed when they entered cold regions. In 1965 a large pressure vessel, intended

Figure 8.3   New thoughts on an old accident.

for use at 350 bar, was pressure tested during the winter and failed catastrophically[29]. One piece weighing 2 tonnes went through the workshop wall and travelled nearly 50 m. Pressure testing should be carried out above the transition temperature, the water being warmed if necessary. Bursting discs have ruptured because they were too cold. At the other end of the scale excessive temperature has caused the failure, by creep, of many furnace tubes.

The accident to Icarus illustrates the hazards of taking equipment outside the limits specified by the designer. Icarus was warned but chose to ignore the design limitations, with fatal results. Let other operators be warned by his misfortune. And designers should, like Daedelus, make sure that the operating staff are fully aware of the limitations of the equipment they are using.

REFERENCES IN CHAPTER 8
1. Quoted by Epstein, L.C., 1981, *Relativity Visualised* (Insight Press, San Francisco, USA), 100.
2. Stern, G.B., 1944, *Benefits Forgot*.
3. Hilton, M. and Marshall, G., 1988, *The Gospels and Rabbinic Judaism* (SCM Press, London, UK), 82.
4. Kletz, T.A., 1988, *Learning from Accidents in Industry* (Butterworths, London, UK).

5. Egginton, J., 1980, *Bitter Harvest* (Secker and Warburg, London, UK).
6. Cullen, W.D., 1990, *The Public Inquiry into the Piper Alpha Disaster* (HMSO, London, UK), Paragraph 14.51.
7. Dept of Transport, 1987, *MV Herald of Free Enterprise: Report of Court No 8074: Formal Investigation* (HMSO, London, UK).
8. *British Medical Journal*, 301, 4 August 1990, 300.
9. Health and Safety Executive, March 1977, *The Explosion at the Dow Chemical Factory, King's Lynn, 27 June 1976* (HMSO, UK).
10. Stewart, R.M., 1971, High integrity protective systems, *Symposium Series No 34* (Institution of Chemical Engineers, Rugby, UK), 99.
11. Pickles, R.G., 1971, Hazard reduction in the formaldehyde process, *Symposium Series No 34* (Institution of Chemical Engineers, Rugby, UK), 57.
12. Gillett, J.E., February 1985, *Process Engineering*, 66 (2): 19.
13. Kletz, T.A., 1992, *Hazop and Hazan — Identifying and Assessing Process Industry Hazards*, 3rd edition (Institution of Chemical Engineers, Rugby, UK; published in the USA by Taylor & Francis, Bristol, PA), Section 5.2.5.
14. Kletz, T.A., 1988, *Learning from Accidents in Industry* (Butterworths, London, UK), Chapter 16.
15. Bartlett, J.V. (Chairman), July 1986, *The Accidental Flooding at the Victoria and Albert Museum on 21/22 March 1986* (Property Services Agency, London, UK).
16. Murphy-O'Conner, J., 1986, *The Holy Land*, 2nd edition (Oxford University Press, UK), 159.
17. *Annual Report of Her Majesty's Inspectors of Explosives for 1970* (HMSO, London, UK), 19.
18. White, S., private communication.
19. Petrowski, H., 1982, *To Engineer is Human* (St Martin's Press, New York, USA), Chapter 8.
20. Anderson, R.C., *The Great Orme Railway*.
21. *Report on the Accident with the Linear Accelerator at the University Clinical Hospital of Zaragoza in December 1990*, 1991, Translation No 91–11401 (8498e/813e) (International Atomic Energy Agency, Vienna, Austria).
22. *Health and Safety at Work*, November 1991, 13 (11): 10.
23. Spooner, P., 1992, *Disaster Prevention and Management*, 1 (2): 28.
24. Sorrell, M., 26 August 1992, *Never Mind, I Stopped my Train* (BBC Radio 4).
25. Currie, J.R.L., 1971, *The Runaway Train, Armagh 1889* (David and Charles, Newton Abbott, UK).
26. Jones, M.W., 1976, *Deadline Disaster — A Newspaper History* (David and Charles, Newton Abbot, UK), 20.
27. Curzon, R., 1849, *Visits to Monasteries in the Levant*, quoted by Osband, L., 1989, *Famous Travellers to the Holy Land* (Prion, London, UK), 36.
28. Quoted by H.Petrovski in *To Engineer is Human*, 1986, (St Martin's Press, New York, USA), Chapter 14.
29. *British Welding Research Association Bulletin*, June 1966, 7 (6): 149.

9. CHANGES IN SAFETY — A PERSONAL VIEW

'... when it is impossible to write an account of what men know, write an account of what they have to learn.'
Denis Diderot (1713–1784)[1]

An account of my personal experiences will show how the resources devoted to safety and the profile of the subject have changed over the years, and how the views described in this and earlier books were developed.

9.1 EARLY DAYS

I joined ICI at Billingham in the North East of England in 1944, at the age of 21, after graduating in chemistry at Liverpool University. Like most of the young chemists engaged at the time I started in the research department and then — in 1952, nearly eight years later — I was transferred to the works as a plant manager (the equivalent of a supervisor in most US companies). I received no formal training in safety or in the responsibilities of a manager, but I spent several weeks on shift with the shift managers and a few weeks on days with my predecessor. I was placed in charge of a unit which manufactured *iso*-octane from butane in a three-stage process:

(1) Dehydrogenation to butene;
(2) Dimerisation to *iso*-octene;
(3) Hydrogenation to *iso*-octane.

The second and third steps were mild reactions but in the first step butane at 570°C and 5 bar gauge was passed through catalyst tubes in a furnace. The catalyst had to be regenerated every hour by burning off carbon so there were two sets of tubes. Every hour the operating set was swept out with nitrogen so that the catalyst could be regenerated while production was continued in the other set. The process was thus a hazardous one but in fact operated remarkably smoothly, and neither I nor anyone else thought of it as hazardous. At the end of 1953 or early in 1954 a unit for the manufacture of acetone from isopropanol was added to my responsibilities.

Two incidents which occurred during this period taught me useful lessons and may have contributed to my future interest in safety.

One was the first lost-time accident in which I was involved. A member of the research department wanted to attach an instrument to a branch on a caustic soda pipeline and carry out some measurements. The foreman and I thought that the instrument looked too fragile to withstand the pressure but the research worker assured us that he had used it before. I let him go ahead but decided to watch, with the foreman, from a safe distance. The instrument leaked, spraying caustic soda over a visiting workman who, like the postman in one of G.K. Chesterton's Father Brown stories, had not been noticed by the foreman or myself. He was washed at once and was not seriously burnt but had an allergic reaction a day or two later and was absent for a few days.

My boss wrote the report on the accident and blamed the research worker but I felt responsible; I believed, and still do, that a manager is responsible for everything that goes on on his plant. The incident taught me that a manager should not be afraid to back his own judgement against the expert's, especially when the expert says it is safe and the manager is doubtful. Obviously expert advice should not be disregarded lightly, especially when the reputation of the expert is high, but one of the skills a manager has to acquire is the ability to assess the mass of expert and often conflicting advice that he constantly receives.

The incident also taught me that I should not have simply accepted the research worker's word that the equipment had been used before in similar situations: I should have questioned him about the precise conditions of use.

The second incident involved a pumphouse. Any spillages drained out into an open trench through pipes fitted with U-bends. The U-bends did not seem to fulfil any useful purpose, and rubbish collected in them, so I had them removed. One of the shift foremen was on holiday at the time. When he returned he told me that the U-bends had been installed about ten years earlier following a fatal accident. A leak of flammable gas had entered the pumphouse through the drain lines and had exploded. The U-bends had been installed so that, filled with water, they would prevent this happening again. The other foremen either never knew about this explosion or had forgotten.

During the intervening ten years the pumphouse walls had been removed, it became no more than a roof supported on four poles, so I was right, the U-bends no longer served any purpose. But nevertheless the incident taught me a useful lesson: *Never remove anything, equipment or procedure, unless you know why it was put there in the first place.*

These two incidents were not disasters but we do not have to wait until disasters occur to learn valuable lessons. The still small voice of minor incidents can teach us as much as thunder and lightning.

In 1955 I left the *iso*-octane and acetone plants and moved to a unit that separated various substituted phenols from a tarry by-product produced in another part of the site. Exchanging smooth-running continuous plants for troublesome batch ones was a cultural shock and the 18 months I spent on the plant were the most tiring of my whole career. Safety was more to the fore as the products were corrosive and chemical burns, mostly minor, were common.

About the same time I was appointed the part-time safety officer for the works, an organisation of about a thousand people. I was expected to spend only a few hours a week on this task, looked upon as a chore that had to be carried out by one of the plant managers. I was assisted by an elderly, worn-out foreman who ordered protective clothing, kept the accident statistics, liased with the Billingham Site safety department (the Billingham site consisted of about six works, each with considerable autonomy), attended accident investigations, and toured the works on the lookout for mechanical hazards. I supervised the foreman, attended the investigations of the more serious accidents, introduced a few new items of protective clothing and followed up a few safety problems, mainly mechanical ones, such as poor access. The set-up illustrated the attitude to safety at the time. There was a genuine concern that people should not get hurt but no realisation that the subject required either a technical input or the attentions of an experienced manager. The only measure of safety was the lost-time accident rate. Safety seemed a dull subject, a view I was to change later.

In 1956 I left the works for a two-year spell in the Technical Department. The following year there was a propylene fire on the works and four men were badly burned. I knew two of them quite well and went to see them in hospital. Although I had been warned what to expect I shall never forget the shock of walking into the ward, seeing four blackened faces and not being able to tell who was who. If I go further than some people might consider necessary in recommending precautions against LPG fires, this is the reason. Anyone else who had seen them would do the same. Fortunately they made a complete recovery, physically, but when one of them was retiring early, ten years and two promotions later, he told me that he had never really got over the mental shock.

9.2   CHANGING RESPONSIBILITIES

In 1958 I returned to the works as a section manager (the equivalent of a superintendent in the US) and, in 1961, became one of the two assistant works managers. Safety was, of course, one of my line responsibilities but I took more interest in it than some of my colleagues on other works. The lost-time accident rate was still the only measure of safety; works managers were pilloried by the production director if it was considered too high and so a lot of effort was spent persuading injured men to continue at work. Figure 9.1 (see pages 146 and 147), based on some contemporary cartoons, exaggerates — but not all that much — the attitude of the time.

We held weekly meetings to discuss accidents, however trivial, and the actions we should take. The meetings made me realise that managers could

LESSONS FROM DISASTER

Figure 9.1   How to reduce the lost-time accident rate — a 1950s view.

CHANGES IN SAFETY — A PERSONAL VIEW

usually do something to prevent an accident happening again. ICI's official statistics showed that 60–70% of accidents were due to 'human failing' but this was clearly wrong. I became reluctant to accept 'human failing' as the cause of an accident. I also insisted that dangerous occurrences, in which no-one was hurt, were investigated and reported as thoroughly as accidents. These actions would not be worth noting today but were not common at the time.

One of the more effective actions I took was to go round the works with a camera, photographing the hazards I saw, and then show the slides to a meeting of the technical staff. People were shocked to see hazards that they passed every day without noticing.

The safety adviser on another works tried the same technique but was less successful. When he showed the slides the works manager, who was present, was so shocked that he lost his temper and played merry hell with the people responsible. The safety adviser was not popular.

In 1965/6 there were several fires and explosions on the works and these made me realise that there was more to safety than the subjects usually considered under that heading. I began to take more interest in the reasons why fires and explosions occurred, managerial as well as technical, and the actions needed to prevent them.

One of these fires taught me a number of valuable lessons. When the works was built in 1935, fixed roof tanks containing petrol and other light hydrocarbons were blanketed with nitrogen, a forward-looking decision for the time. After some years the system fell into disuse but was reinstated in the early 1960s. The design of the system was unusual: the vapour spaces of all the storage tanks were connected to a nitrogen gasholder. If the level in a tank rose, nitrogen was pushed into the holder; if the level fell, nitrogen was drawn from it. This conserved nitrogen which was expensive at the time.

Despite the nitrogen blanketing, an explosion occurred in a storage tank, the roof was blown off and the contents were set on fire. Afterwards we found that the tank was isolated from the blanketing system; the valve in the connecting line leading to the nitrogen gasholder was closed. Seven months before the explosion the plant manager had checked that the valve was open but no-one had looked at it since then.

We were fortunate that the valve was closed as the 'nitrogen' in the gasholder contained 15% oxygen! There were leaks in the system and no regular analyses. If the tank had been connected to the gasholder the explosion might have spread to all the other tanks that were connected.

This incident made me realise that all protective equipment should be checked regularly. The equipment was not the most modern but adequate if used correctly, yet when I discuss this incident with groups of managers and engineers (in the manner described in Section 10.2, page 167) they often want to install more and better protective equipment. But if people do not use the equipment they have got, will they use new equipment? We need to ask why it was not used. The incident made me realise that the training was poor. We had not convinced the foremen and operators that nitrogen blanketing was necessary. They had managed without for it decades and saw no reason to change. I began to realise that it is not enough to issue safety instructions. We have to explain the reasons for them and check up to see that they are being followed.

The source of ignition was never found with certainty but possible sources, and the accident as a whole, are discussed in Reference 2.

The lessons of the tank fire were reinforced a few weeks later when an explosion occurred in a flare stack. It was used intermittently and there was supposed to be a flow of nitrogen up the stack to prevent air diffusing down and to sweep away any small leaks of air into the stack. The nitrogen flow had been reduced almost to zero, to save nitrogen. The flow was not measured and the atmosphere in the stack was never analysed.

As a plant manager I was always looking in eyewash bottle cabinets, trying out safety showers and so on and I continued to do so as a section and assistant works manager. (Even today, if I visit a plant, I have to restrain my desire to poke and pry in the corners.) The two explosions showed me that I would have to extend my curiosity to a whole range of more technical safety matters. Many could not, of course, be checked visually and I would have to ask for analysis results, inspection reports and so on.

In the other works of ICI Petrochemicals Division there were a number of serious fires in 1965/7 and a number of people were killed. The Board decided that there was a need for a technical input to safety; my interest in the subject was known and I was asked to write a job remit for a full time technical safety adviser. I described the experience necessary, a 'coat' which fitted me exactly, and started the new job in January 1968. I expected to change jobs again after a few years, in the normal ICI mid-career way, but I was to stay in safety until I retired in 1982, and to continue working in the field thereafter. At the time the appointment of someone with my experience and seniority to a full time job in safety was unusual; if the reason for the appointment had not been so obvious I would have wondered what I had done wrong.

9.3   SAFETY ADVISER
My remit was a broad one: to advise design and operating staff on the action they should take to avoid accidents and dangerous occurrences. I set for myself several guidelines which I was to follow for the rest of my ICI career:

(1) I decided to concentrate on technical accidents, mainly fires and explosions, rather than mechanical accidents. Although the latter were the most numerous, technical accidents caused half the deaths, the most serious injuries and most of the material damage and lost production. As a result I paid little attention to the lost-time accident rate. Also, the traditional safety officers were still around and could look after the cuts and bruises. They did not resent an outsider entering their territory; they felt, rightly, that I was doing a different job and in any case they were not the sort of people who would have been resentful.

(2) I tried to keep a sense of balance, concentrating on the most probable and serious hazards and advising people not to spend time and money dealing with unlikely and trivial ones. This led to my interest in hazard analysis, or quantitative risk assessment as it is now often called[3].

(3) I tried to sell my ideas softly but persistently. I had no authority to impose them nor did I wish for any. A manager must remain in control of his plant. I tried to convince managers and designers, by technical arguments and by describing accidents that had happened, that they should incorporate certain ideas or practices into their plants. To get my ideas across I held regular discussions of accidents that had happened, described in the next chapter, and started a monthly *Safety Newsletter* that ultimately achieved a very wide circulation within the company and outside.

At first I was a lone worker, but as I built up a small team I emphasised to them that a job was not finished when they made their recommendations, only when the manager or designer concerned had carried them out, or had convinced them that they were wrong.

(4) Whenever possible, if I had an idea, I just got on with it, rather than asking for permission or agreement. I did not ask for permission to start a monthly *Safety Newsletter* or weekly discussions or if I should set numerical criteria for safety. I did not ask the Board, for example, to agree that flammable gas detectors should be widely installed for the detection of leaks. Instead I persuaded individual managers or designers that it was the right solution for their particular problem and gradually it became the accepted practice, the common law of the Division. I could, of course, follow this policy only because it was encouraged by the ICI culture and because everyone knew that the Division Board was broadly supportive of my ideas.

The ability to be able to get on with something, rather than be forever lobbying one's bosses for agreement, was of tremendous advantage. To quote my former colleague, Derek Birchall:

'The reason for the speed and success of the industrial revolution in this country was largely the fact that many of the inventors did not need to seek support, but simply got on with it. Abraham Darby did not have to write a submission in order to attempt the experiment of using coke rather than charcoal in iron smelting — he was master of his own experiment and of its consequences.'[4]

Or, to take a more recent example, would Pilkingtons have been able to develop their ultimately very successful float glass process if they had been a public company at the time rather than a family business?

I entered the safety field without any training or full time experience of the subject and in some ways this was an advantage: I avoided the restricting influence of traditional ways of thinking. I did not question the ways of the traditional safety officer or deliberately ignore them; I was simply unaware of them. I was a bus rather than a tram. Today there are many courses for safety professionals, on both the technical and managerial aspects of the job, and on balance they make people into better safety advisers, but there is a down side: the courses can produce clones of those who devise them and thus perpetuate traditional ways of thinking.

I started out as a one-man band but soon realised that I needed a colleague who could spend his time on the plant seeing whether or not safe practices were being followed and giving advice when they were not. The first safety surveyor to join me made a detailed survey of static electricity hazards and then, joined by a colleague, looked at permits-to-work, trip testing, open vents, LPG equipment and much more. They carried out surveys of specific topics, as described in Section 7.4 (page 96), rather than audits of everything. Ultimately we became a group of seven or eight people, some of them transfers from other departments who brought their work with them. Thus responsibility for hazard and operability studies was transferred, along with Herbert Lawley, from the Management Service's Department to Safety and Loss Prevention Group, as we were called.

In 1979, after ten years in the job, I published a paper on 'A decade of safety lessons'[5]. It described the themes which, looking back, had dominated each year's activities, particularly my regular weekly safety discussions (see Section 10.2, page 167) but also our auditing and other activities. The paper (updated to the time of my retirement) is summarised below to give an indication of the technical problems that dominated our thinking. Many of them are discussed in greater detail elsewhere in this book.

LESSONS FROM DISASTER

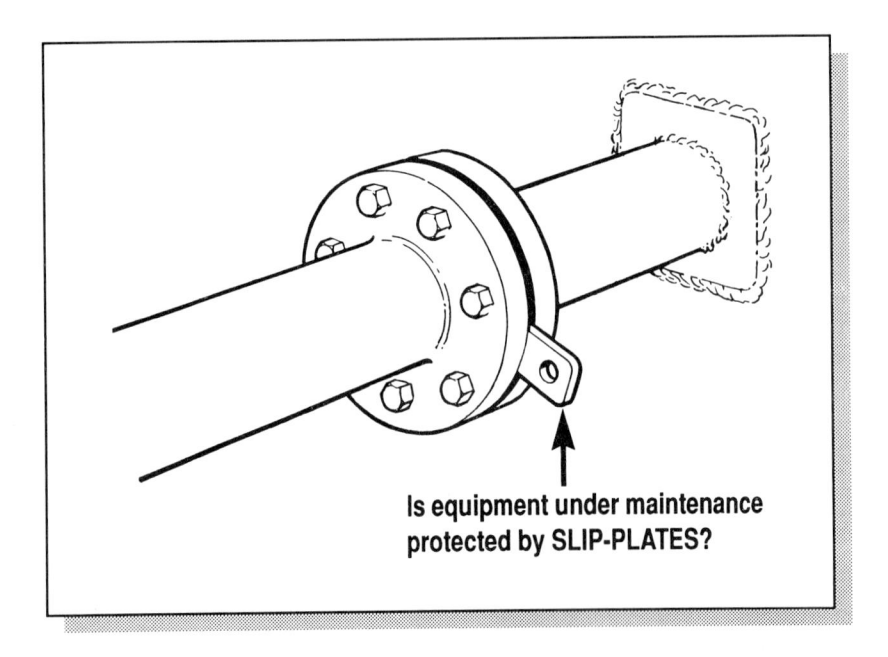

Is equipment under maintenance protected by SLIP-PLATES?

Figure 9.2   1968 — the year of the slip-plate.

1968 — THE YEAR OF THE SLIP-PLATE (Figure 9.2)
The incident that, more than any other, led to my appointment as safety adviser, occurred because a pump was opened for repair with the suction valve open (see Sections 2.1 (page 4), 6.1 (page 70) and 7.8 (page 109)). New procedures for isolating equipment before maintenance had already been introduced but I knew that these would lapse, as they had done before, unless people were convinced that they were necessary and regular checks were carried out to make sure that they were followed. I looked into the identification of equipment as well as its isolation — a number of accidents had occurred because the wrong item had been opened up — and persuaded the works to use numbered tags to identify equipment which had to be repaired (Figure 9.3).

1969 — THE YEAR OF THE BUILDING WITHOUT WALLS (Figure 9.4)
About a year after my appointment two serious incidents, a fire and an explosion, occurred on ICI's Wilton site, just across the River Tees from Billingham. Although the plants belonged to other Divisions I was asked to join the investigating teams and learned a lot by doing so. The explosion was due to a leak into

CHANGES IN SAFETY — A PERSONAL VIEW

Tag along with us

147

They make the job exact and neat...
Please return when job complete.

Figure 9.3   1968 — we emphasised the identification of equipment under repair.

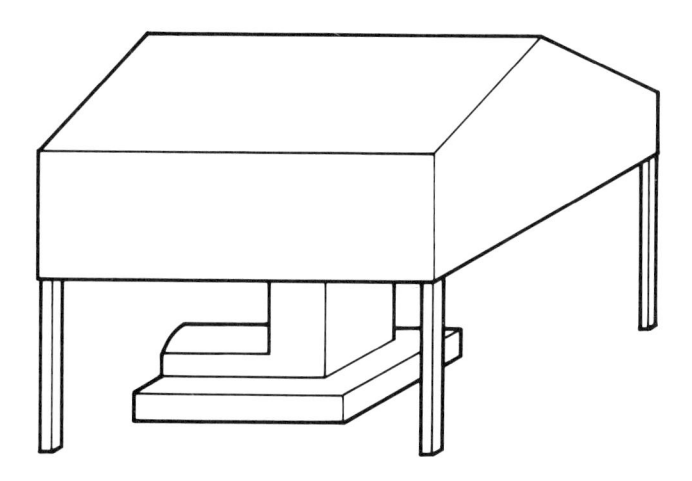

Figure 9.4   1969 — the year of the building without walls.

153

LESSONS FROM DISASTER

a confined space and could have been avoided by better ventilation. On the whole Petrochemicals Division's plants were built in the open air but there were some exceptions and we set about improving these (see Section 2.2, page 10). In particular the walls of a new compressor house were pulled down as soon as they were built.

Also emphasised during this year was the need for automatic detection of leaks and their isolation by remotely-operated emergency isolation valves.

1970 — THE YEAR OF THE RUPTURED VESSEL
The end of a tank was blown off, killing two men who were working nearby. The tank was used to store a liquid of melting point 100°C and was heated by a steam coil. The tank inlet line was being blown with compressed air to prove that it was clear, the usual procedure before filling the tank, but the vent was choked. It was known to be liable to choke but this was looked upon as an inconvenience rather than a hazard. No-one realised that the air pressure (5.5 bar gauge or 80 psig) could burst the tank (designed for 0.35 bar gauge or 5 psig). Some time in the past the vent size had been reduced from 6 inches (150 mm) diameter to 3 inches (75 mm) (see Section 2.3, Incident 2 (page 16)). The works had to be persuaded that open vents are relief valves and should be treated like relief valves: inspected regularly, heated if necessary and their size not changed without checking that the new size is adequate. One of the safety surveyors set out to personally inspect every open vent in the Division.

1971 — THE YEAR OF THE BLEVE (Figure 9.5)
A few years earlier a serious BLEVE (Boiling Liquid Expanding Vapour Explosion) in France (see Section 6.3, page 76) had drawn attention to the hazards of LPG. A colleague had drawn up new standards (Table 6.2, pages 78–80) and we carried out a thorough survey of all the Division's LPG installations, making many recommendations for improvement. To convince people that change was necessary, LPG incidents were discussed in the weekly safety meetings. One works was rather reluctant to make any changes. Then they had a small LPG fire of their own (see Section 6.3). It was nothing compared to the French BLEVE and no-one was hurt but it had far more impact. Like waves in a pond, the effects of an accident decrease with distance.

1972 — THE YEAR OF THE TRIP TEST
As discussed in Sections 6.2 (page 74) and 7.5 (page 101), modern plants depend for their safety on trips and alarms which cannot be relied on unless they are tested regularly. The two safety surveyors set out to witness the testing of every

Figure 9.5   1971 — the year of the BLEVE.

trip and alarm in the Division, a tremendous undertaking. The results varied widely. On many large plants the surveyors made only minor recommendations. On some plants that had been transferred into the Division during a reorganisation in 1971, testing was poor or non-existent and many trips were out of order. Since the plants seemed to operate they probably contained more trips than they needed and a number were removed.

Readers will have realised by now that, in my view, it is not sufficient for safety advisers to advocate policies. In addition, we should be willing to describe in detail precisely what should be done and check that it is being done.

1973 — THE YEAR OF THE FURNACE (Figure 9.6 — page 156)
In 1972 a furnace tube had burst and damage was extensive. The investigation showed that many people did not fully understand the principles of furnace operation and, in particular, that overheating will shorten tube life and that furnaces operate close to failure conditions. A pressure vessel can withstand several times its design pressure without bursting but a furnace tube can withstand only a few percent increase in its absolute temperature. Furnace failures were featured in the weekly safety discussions and in the *Safety Newsletters*.

LESSONS FROM DISASTER

> **An elephant has a good memory...**
>
> **...BUT A FURNACE TUBE HAS A BETTER ONE!**
>
> If you let your furnace run 60°C hotter than design for 6 weeks, you may half the life of the furnace.

Figure 9.6   1973 — the year of the furnace.

1974 — THE YEAR OF FLIXBOROUGH[6-8]

Although the explosion did not occur until halfway through 1974 it so dominated the rest of the year that it became the year of Flixborough. The immediate lessons drawn were:

- Modifications should be controlled (see Figure 9.7).
- All managers, including the most senior, should spend some time walking round the plant, on the lookout for anything unusual.
- Companies should make sure they have enough staff with the right professional qualifications and experience.
- Plants should be laid out so as to avoid knock-on or domino effects.
- Occupied buildings close to hazardous plants should be blast-resistant.

1975 — THE YEAR OF THE MOD

As the dust of Flixborough settled it became clear that modifications (mods) of all sorts had to be controlled much more closely than in the past. Other, local, incidents brought home the same message and it became the theme of the year (see Figure 9.8 and Section 7.8, page 111). One of the side-effects was that we made fewer but better modifications.

CHANGES IN SAFETY — A PERSONAL VIEW

"Any modifications should be designed, constructed, tested and maintained to the same standards as the original plant."

From the official report on the explosion at Flixborough on June 1st 1974

Figure 9.7   1974 — the year of Flixborough.

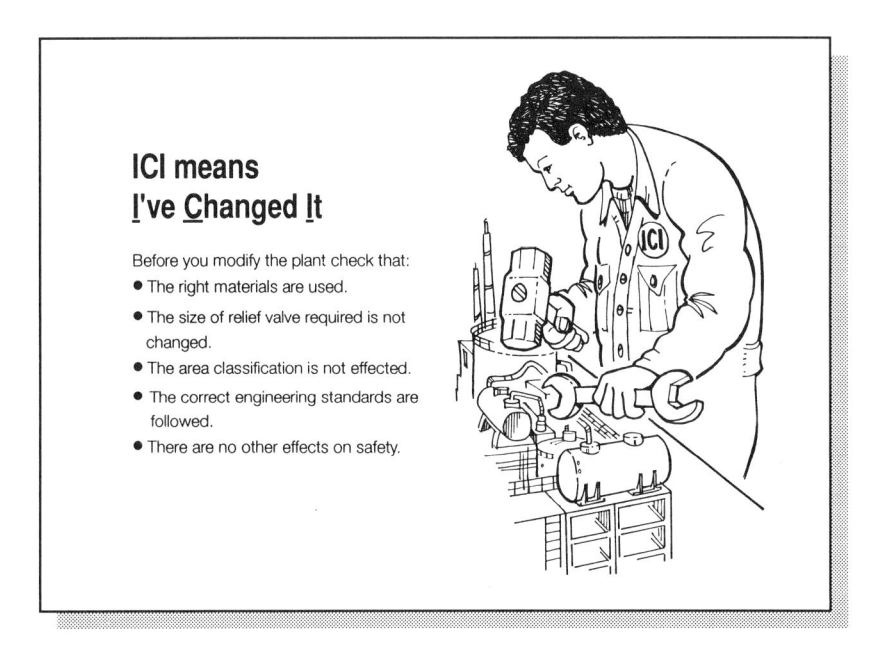

ICI means
I've Changed It

Before you modify the plant check that:
- The right materials are used.
- The size of relief valve required is not changed.
- The area classification is not effected.
- The correct engineering standards are followed.
- There are no other effects on safety.

Figure 9.8   1975 — the year of the mod (that is, modification).

157

If you...

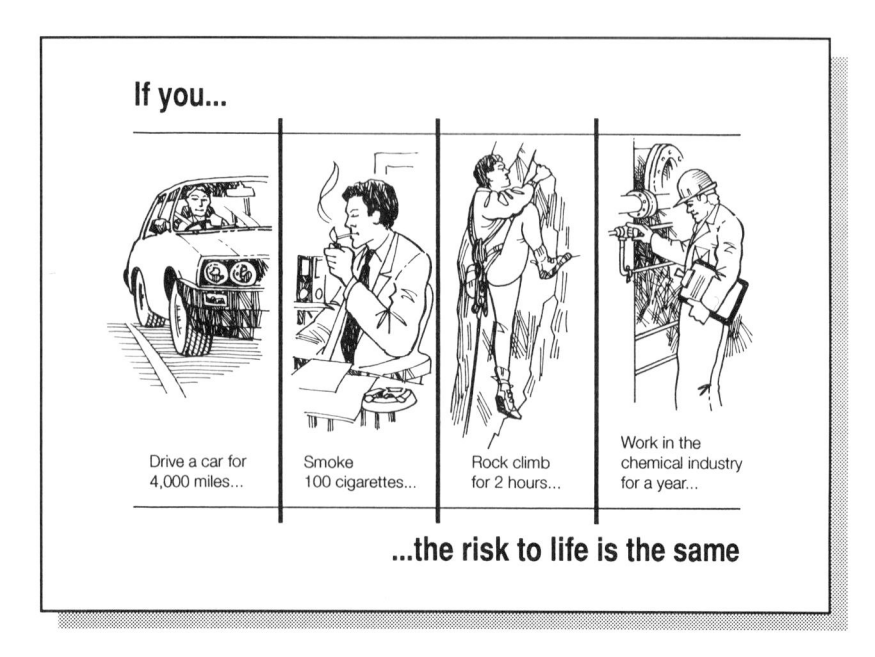

| Drive a car for 4,000 miles... | Smoke 100 cigarettes... | Rock climb for 2 hours... | Work in the chemical industry for a year... |

...the risk to life is the same

Figure 9.9   1976 — the year of the sum.

1976 — THE YEAR OF THE SUM

Since the late 1960s ICI had pioneered the application of numerical methods to safety problems (see Section 6.4, page 81). We cannot do everything at once so we need a rational method to help us decide which risks should be dealt with first and which left, at least for the time being (Figure 9.9). By 1976 people were saying, 'Don't tell us why, tell us how' and we devoted much of our safety training to this subject. I think I learnt as much as the students. Unless one has the opportunity to explain one's ideas to others, they inevitably remain vague and ill-defined. The notes I used formed the basis of my book on *Hazop and Hazan*[3].

1977 — THE YEAR OF THE MAN IN THE MIDDLE

A number of accidents had occurred because someone forgot to close a valve or pressed the wrong button. The alarm had worked and the valves and buttons were in order but the man in the middle did not carry out his part of the task correctly. We realised that it was little use telling him to be more careful and that we had to accept an occasional error or change the work situation — the design or method of working — so as to remove or reduce opportunities for error (see Section 7.7, page 108). We discussed a number of such accidents at the weekly safety meetings. I hoped there would be a certain amount of conflict between managers, who wanted design changes to prevent errors, and design engineers who felt that we already went far enough down that road. There was no conflict. The design engineers had all worked in production and already realised the need to remove opportunities for error (or protect against the consequences of error or provide opportunities for recovery).

1978 — THE YEAR OF THE LOWER INVENTORY (Figure 9.10)
Many plants contain a large amount of hazardous materials. We try to keep it under control by adding on trips and alarms and other protective equipment which may fail or can be neglected. It would be better to devise processes and equipment that use less hazardous material, so that it does not matter if it all leaks out, or safer materials instead. During 1978 I tried to draw attention to ways in which this could be done (see Section 6.5, page 83). My first paper on inherently safer design, as it is now called, was published[9], following a major company seminar on the subject at the end of 1976.

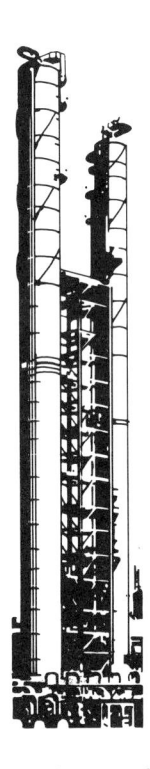

Figure 9.10 1978 — the year of the lower inventory. Is this at the frontiers of technology or is it a dinosaur?

1979 — THE YEAR OF THE SIMPLER DESIGN (Figure 9.11 — page 160)
Extending the theme of the previous year, I advocated simpler designs as greater complexity means more opportunities for equipment failure or human error. We held a Division seminar on the subject.

LESSONS FROM DISASTER

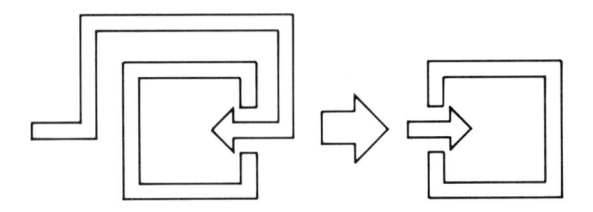

Figure 9.11   1979 — the year of the simpler design.

1980 — THE YEAR OF THE PAST
The theme of the year was the theme of this book: the need to remember the lessons of the past.

1981 — THE YEAR OF THREE MILE ISLAND
The accident at the nuclear power station at Three Mile Island occurred in 1979. By 1981 much information was available and we emphasised the lessons for the chemical industry. There were many[10], including:
- Give ancillary equipment as much attention as the heartland.
- Do not use instrument air for line blowing.
- Measure directly what we need to know.
- Train operators in fault diagnosis so that they can handle problems not foreseen when the instructions were written.
- Learn the lessons of incidents elsewhere.
- Plan ahead how to handle public relations in an emergency.
- Design inherently safer plants which are not dependent, or are less dependent, on added-on safety systems which may fail or may be negelected.

The last item was, of course, an extension of the 1978 theme. Inherently safer designs of nuclear reactor are available although they have never been taken beyond the development stage. In the UK the decision to build a pressurised water reactor at Sizewell was in some ways a retrograde step as the gas-cooled reactors previously built are less dependent on added-on safety systems which may fail or be neglected[11]. Of course, the UK operating company, Nuclear Electric, is a highly professional organisation and I have no doubt that they can be relied on to maintain the protective systems in full working order. I would have said the same about Dutch State Mines before the accident at Flixborough in 1974 and about Union Carbide before the accident at Bhopal in 1984. (However, Flixborough and Bhopal were both peripheral overseas plants. Sizewell is Nuclear Electric's flagship.)

1982 — NOT THE YEAR OF THE AUDIT
For various reasons our auditing and surveying effort had lapsed. During the last few months before I retired from ICI I was emphasising the importance of getting it restarted.

I do not want to give the impression that the theme of each year was planned in advance. The pattern I have described was seen looking back, not looking ahead. To quote Robert Jungk, 'What was seen in retrospect as a difficult but straight road leading to its goal was really a labyrinth of winding streets and blind alleys'[12]. Several important themes such as hazard and operability studies are not mentioned because there was no year in which they received particular attention.

It may be of interest to look back over the years since I left ICI and note briefly the highlights of each year. I was, of course, now trying to respond to the needs of a wider circle and not just those of ICI Petrochemicals Division.

1983 — THE YEAR OF PRACTICAL ADVICE
My main activity was extracting incidents from my *Safety Newsletters* for publication in my first book, *What went wrong? — Case Histories of Process Plant Disasters*[13]. While some of my later books deal with ideas, this one is a collection of practical advice on ways of preventing accidents on process plants, based on incident reports (most of them were not really disasters) from ICI and other companies. It has sold more copies than any of my other books.

1984 — THE YEAR OF BHOPAL (Figure 9.12 — page 162)
This was the year of Bhopal and, like many others, I tried to find out what had happened and what were the lessons for the rest of the chemical industry. There were many, but the most important, overlooked by many commentators, is the need to reduce inventories of hazardous materials (see comments in Section 7.9, page 114). As I have already pointed out, the chemical which leaked and killed over 2000 people was not a product or raw material but an intermediate. It was

Figure 9.12 1984 — the year of Bhopal.

convenient to store it but not essential to do so. Since 1984 many companies have followed the advice given after Flixborough, ten years earlier, and reduced their stocks of hazardous intermediates.

1985 — A YEAR ON WHEELS (Figure 9.13)

I resumed an earlier interest in the transportation of hazardous chemicals[14]. The United Kingdom's record is outstandingly good: on average the transport of chemicals and petrol has killed one person per year. This is one too many, of course, but it has to be considered in the context that 4000–5000 people are killed every year by ordinary road accidents. Chemicals have killed many more in other countries and while the UK's good record is due in part to high standards it may also be due, in part, to good luck. We cannot therefore relax, and public concern will ensure that we do not. The Health and Safety Executive give a good deal of attention to the risks and in 1991 they published a lengthy quantitative study of them[23].

1986 — THE YEAR OF CHERNOBYL

Just as earlier years had been dominated by Flixborough, Three Mile Island and Bhopal, this year was dominated by Chernobyl and its lessons for the chemical industry[15]. The most important lesson, and again one overlooked by most commentators, is the need to encourage the development of inherently safer designs, less dependent on added-on safety systems which may fail, be neglected or, as at Chernobyl, deliberately switched off (see comments on 1981).

1987 — THE YEAR OF PEELING THE ONION
I completed a book on the investigation of accidents, major and minor, and the need to look below the surface for ways of avoiding the hazards and for weaknesses in the management system[2]. It discusses more fully some of the accidents described briefly on earlier pages such as Flixborough, Bhopal, Three Mile Island, Chernobyl and Aberfan as well as others that did not hit the headlines.

1988 — THE USER-FRIENDLY YEAR
I widened the concept of inherently safer design into that of the user-friendly plant, one which can withstand human error or equipment failure without serious effects on safety, output and efficiency[16]. I described the concept and many examples of what has been or could be done. I also completed a new edition of a book originally called *Myths of the Chemical Industry or 44 Things a Chemical Engineer Ought NOT to Know*. The number grew to 60[17]. These two books have not sold as well as *What Went Wrong?*, perhaps because both are primarily ideas books in which I have tried to shatter some misconceptions and, particularly in

Figure 9.13   1985 — a year on wheels.

the first one, suggested some changes in the design process (though they also contain some advice on day-to-day problems).

1989 — THE YEAR OF THE 400
I completed a book of nearly 400 short items on various aspects of safety and loss prevention, technical and managerial, which are not covered in my other books[18]. It is based on hundreds of jottings that I made during nearly twenty years' work in process safety.

1990 — THE YEAR OF HUMAN ERROR
I completed a revised edition of a book on human error, originally published in 1985[19]. It expands the ideas summarised in Section 7.7, page 108.

1991 — THE YEAR OF COMPUTER CONTROL
A paper on accidents on computer-controlled plants[20], an update of one originally presented in 1982, produced several requests for similar papers[21,22]. While I did discuss hardware failures and software errors I was more concerned with incidents that have occurred because the operating team did not fully understand the computer logic, that is, they did not know precisely what it could and could not do, and with weaknesses in the computer/operator interface.

1992 — THE YEAR OF THE RECURRING ACCIDENT
I was working on this book. I also prepared a paper on changes made with the commendable intention of improving the environment which have had unforeseen and undesired effects on safety. The examples discussed include the boxing in of compressor houses to reduce noise (see Section 2.2, page 10), the hasty replacement of CFCs in aerosols by flammable hydrocarbons and explosions in vent collection systems[24].

REFERENCES IN CHAPTER 9
1. Diderot, D. (freely translated), *Perspectives*, quoted by Downs, R.B., 1975, *Famous Books* (Littlefield Adams, Totowa, New Jersey, USA), 120.
2. Kletz, T.A., 1988, *Learning from Accidents in Industry* (Butterworths, London, UK), Chapter 6.
3. Kletz, T.A., 1992, *Hazop and Hazan — Identifying and Assessing Process Industry Hazards*, 3rd edition (Institution of Chemical Engineers, Rugby, UK; published in the USA by Taylor & Francis, Bristol, PA).
4. Birchall, D., *Chemistry and Industry*, 13 July 1983, 539.
5. Kletz, T.A., June 1979, *Hydrocarbon Processing*, 58 (6): 195.

6. Kletz, T.A., 1988, *Learning from Accidents in Industry* (Butterworths, London, UK), Chapter 8.
7. Lees, F.P., 1980, *Loss Prevention in the Process Industries* (Butterworths, London, UK), Appendix 1.
8. Parker, R.J. (Chairman), 1975, *The Flixborough Cyclohexane Disaster* (HMSO, London, UK).
9. Kletz, T.A., 6 May 1978, *Chemistry and Industry*, No 9, 287.
10. Kletz, T.A., 1988, *Learning from Accidents in Industry* (Butterworths, London, UK), Chapter 11.
11. Kletz, T.A., *The Chemical Engineer*, No 427, July/Aug 1986, 63 and No 495, 25 April 1991, 21.
12. Jungk, R., 1960, *Brighter than a Thousand Suns* (Penguin Books, London, UK), 107.
13. Kletz, T.A., 1988, *What Went Wrong? — Case Histories of Process Plant Disasters*, 2nd edition (Gulf, Houston, Texas, USA).
14. Kletz, T.A., July 1986, *Plant/Operations Progress*, 5 (3): 160.
15. Kletz, T.A., 1988, *Learning from Accidents in Industry* (Butterworths, London, UK), Chapter 12.
16. Kletz, T.A., 1991, *Plant Design for Safety — A User-Friendly Approach* (Hemisphere, New York, USA).
17. Kletz, T.A., 1990, *Improving Chemical Industry Practices — A New Look at Old Myths of the Chemical Industry* (Hemisphere, New York, USA).
18. Kletz, T.A., 1990, *Critical Aspects of Safety and Loss Prevention* (Butterworths, London, UK).
19. Kletz, T.A., 1991, *An Engineer's View of Human Error*, 2nd edition (Institution of Chemical Engineers, Rugby, UK).
20. Kletz, T.A., January 1991, *Plant/Operations Progress*, 10 (1): 17.
21. Kletz, T.A., March 1991, *Journal of Process Control*, 1 (2): 111.
22. Kletz, T.A., 1992, *Reliability Engineering and System Safety*, 39 (3): 257.
23. Health and Safety Executive, 1991, *Major Hazard Aspects of the Transport of Dangerous Substances* (HMSO, London, UK).
24. Kletz, T.A., March 1993, Unforeseen side-effects of improving the environment, *American Institute of Chemical Engineers Loss Prevention Symposium*.

10. IMPROVING THE CORPORATE MEMORY

'Memory, of all the powers of the mind, is the most delicate and frail.'
Ben Jonson (1573?–1637), *Explorata: Memoria*

'One reason why history rarely repeats itself among historically conscious people is that the dramatis personae are aware at the second performance of the denouement of the first, and their action is affected by that knowledge.'
E.H. Carr[1]

Looking through a book of quotations for a suitable one to put at the head of this chapter, I found that almost everyone quoted in the section on memory admitted that their memory was poor (though, as La Rochefoucauld says, no-one admitted their judgement was poor) but most of them looked upon this as an advantage. The writer Scholem Asch, for example, wrote, 'Not the power to remember, but its very opposite, the power to forget, is a necessary condition for our existence'.

I have tried to show that in safety the power to remember is vital. So in this chapter I bring together and discuss further the recommendations made in earlier ones, particularly Chapter 2, on the *leitmotiv* of this book: the need for organisations to learn and remember the lessons of the past. I shall not repeat the recommendations made in Chapter 7 on the management of safety.

10.1   SPREADING THE MESSAGE

The first step in remembering a message is to make it available. In Chapter 5 I explain why we should publish our accident reports, discussed the reasons why we do not always do so and made some suggestions. Some companies do not even circulate accident and dangerous occurrence reports widely within the company, restricting the circulation to the minimum number of people who they think need to know. Unfortunately we do not know what we need to know until we know what there is to know.

Suppose a tank is sucked in because the flame arrestor in the vent line is choked; it has not been cleaned for several years although scheduled for cleaning every three months (as in Section 3.1, page 27). The report is sent to all the works managers in the company. Their responses will vary. One may circulate it widely, another may feel so confident that the incident could not occur in his works that he does nothing, a third may carry out a few spot checks or ask someone else to do so, another may intend to do this but never get round to it. If the report is sent direct to every manager and foreman and placed in every control room, then for a few days at least almost everyone in the works is looking at tank vents with sceptical eyes. Even if a works has no flame arrestors, the report may remind them that all protective equipment should be tested or maintained at regular intervals, and that this can lapse once managers lose interest.

However, if everyone is not to be flooded with paper, accident reports need editing. Most of them contain much that is only of local interest (such as, 'At 3.00 pm on 22 July Mr J. Smith, process operator on A shift, was proceeding in a northerly direction along the west side of Building 4 when ...'). Safety advisers should pick out the essential messages, rewrite them if necessary in simpler English and circulate them in periodic newsletters. These should be clearly printed in large print on good quality paper so that they can be easily read by busy people in odd moments. Your company's sales literature will show the standard that is needed (though colour printing is not necessary). The safety adviser also has something to sell.

Whenever possible the actual location of an incident should not be disclosed. This is easy in a large company but may not be possible in a small one where everyone will probably know the location already.

The staff of company libraries and information departments are educated to believe that information is something to be guarded rather than distributed. Their normal reaction, the default action in computer language, is: if in doubt, don't send it out. It is, of course, necessary to restrict the circulation of much information on know-how, future plans, markets and finance, but information on safety should be freely available. Requests should be welcomed, not discouraged. This may be recognised in theory but not always in practice, as openness goes against the grain.

A novel way of spreading knowledge of an accident is to turn it into a short story, with a detective flavour. This has been tried by Mike Tilley[2]. I remembered far more than if I had read the usual report.

10.2   DISCUSSIONS ARE BETTER THAN LECTURES

A talk may convey the message of an accident better than the written word, especially if it is illustrated by slides of the damage. The recommendations can be explained and the audience can comment on them and point out any problems. I used to give such talks, many years ago, when I was a manager, not a safety adviser. One day the training manager said, 'Instead of telling people what happened and what they ought to do, why don't you ask them to tell you?' From this chance remark my colleagues and I developed a training method that we used for a half day almost every week throughout my fourteen years as a safety adviser and which continued after I retired.

The usual participants were a group of 12–20 managers and engineers from the works and from the engineering design, technical and research departments, varying from new graduates to heads of department. Senior foremen also attended from time to time. Nobody had to come; they came because they thought the time would be well spent. The size of the group was important. If more than twenty people were present the quieter ones did not get much opportunity to contribute to the discussions; if less than twelve were present the group was not always critical — in the atomic energy sense — and the discussion did not always take off.

Some of the works safety advisers arranged similar discussions for their foremen and process operators.

I started by describing the technique. Those present were the accident investigation team, a rather large one, and I was all the people they might wish to question rolled into one: the people on the job, the manager, the designer, the technical experts. I then briefly described an accident and illustrated it, whenever possible, by slides of the damage and, if necessary, diagrams of the plant or equipment. The group then questioned me to establish the facts that they thought important, that they wanted to know. I answered questions truthfully but, as usually happens in accident investigations, I did not answer questions I was not asked.

After they had established the facts the group went on to say what *they thought* ought to be done to prevent it happening again. Because the group had developed the recommendations they were more committed to them than if they had merely been told what they ought to do. Sometimes everyone agreed on the actions; at other times there was a lively discussion.

Compared with a lecture, discussions are slow, but more is remembered. We usually spent at least an hour, perhaps half the morning, on a major incident such as one of those described in Chapter 2, although all the information conveyed could probably be covered (though not remembered) in 20 minutes talking or 5–10 minutes reading. We usually spent the rest of the morning on several less complex incidents and the discussions continued over an informal lunch.

Although the discussions took longer than a lecture they achieved more in the time available than discussions in syndicates. The discussion leader can prevent the discussion following paths that lead nowhere and can stop people riding their hobby-horses for too long.

Each programme ran for the best part of a year, different people coming each week. The following year, we discussed different accidents but some of the old modules (as we called them) were repeated from time to time for the benefit of newcomers. Most of the technical staff attended year after year and the discussions gradually increased their knowledge of safety and changed their attitude towards it.

Most of the modules had a common theme such as:
- Preparation for maintenance;
- Over- and under-pressuring of vessels;
- Fires and explosions;
- Human error;
- Failures of alarms and trips;
- Furnace fires and explosions;
- Plant modifications;
- Major accidents that have happened again.

After each module had run its course I summarised the accident reports and the recommendations in a booklet which was sent to all those who had attended and often to others as well.

Many of the notes and sets of slides that we used have been published by the Institution of Chemical Engineers, UK[3].

Leading (or attending) these discussions is much harder than giving (or listening to) a lecture and I was usually fairly tired by lunchtime. Many people come to a safety meeting expecting a quiet rest and are at first surprised to find that they are expected to contribute. After the discussion leader has outlined the first accident and asked the group to do the detective work and find out why it happened there may be an awkward silence. The discussion leader should sweat it out. In time someone will talk. Some discussion leaders find this difficult, start talking themselves and what should be a discussion degenerates into a lecture.

The discussions vary remarkably, even though all those present come from the same company and culture. One week the group may want to redesign every plant; the next group may prefer to change the procedures. Some groups are interested in the immediate technical causes of an accident; others probe the underlying weaknesses in the management system. The discussion leader can comment on the group's ideas and can suggest points for consideration but should not pursuade the group to go down roads they do not wish to follow or to come to conclusions they do not agree with.

If there are any foremen present, they are often the first to see why the accident happened. They may not understand the theory but they have seen similar incidents before. In contrast, young graduates get there in the end but explore many possible scenarios on the way.

I learn a lot from these discussions. My books and papers are scattered with remarks made during them (including the sub-title of this book) for which I have unfairly received the credit.

Although many of the notes and slides are available, the technique has never really caught on outside the Petrochemicals Division of ICI and its successors. The American Institute of Chemical Engineers has run a short course on chemical plant accidents, based on the modules, for many years but attendance has, on the whole, been small.

10.3 REMEMBERING THE MESSAGE

Spreading the lessons of an accident, though not always done well, is usually done much better than the next step: seeing the lessons are not forgotten. This is often not done at all and by the time those who were around when the accident occurred have moved on or retired, the accident is ready to happen again. Here are some actions we can take:

- Describe old accidents as well as recent ones in safety newsletters.

- Discuss some old accidents as well as recent ones at safety meetings or discuss them at meetings of the type described in Section 10.2.

- Include serious accidents of the past in the training of new recruits.

- Prepare a booklet or series of booklets describing the accidents that influenced the company's policy. The Amoco Oil Company has produced an excellent series of booklets and has made them available to other companies[4].

- On each unit keep a memory book (or black book), a folder of reports on past accidents which is compulsory reading for new recruits and which others dip into from time to time. Do not clutter it up with reports on cuts and bruises but do include reports of interest from other plants and other companies. Keep the book in a readily accessible place such as the control room.

Memory books, like all other safety procedures, will not last unless managers take an interest in them and are seen to look at them from time to time and comment on their contents.

- Never remove equipment or stop following procedures unless you know why they were introduced (see Section 9.1, page 143). Modification control procedures might well include the question, 'When and why was this equipment installed/procedure introduced?'

- Include, in plant audits, a check that these actions are being carried out. Ask the manager, foremen and operators what they know about an accident that occurred several years ago or about a major incident in a competitor's plant (see Section 4.6, page 59). Check that the recommendations made in some recent

reports have been adopted (or, if not, that the reasons have been documented).
- Codes, standards and operating instructions are often changed after an accident. A note or appendix should say why the rule or recommendation was introduced and give a reference to the accident report. If this is not done then in time the reason will be forgotten and someone, rightly keen to reduce costs or avoid unnecessary work, will abandon the rule or recommendation. (There are examples in Chapter 4.)
- Before people with long experience retire, ask them to write down some of their accummulated wisdom and knowledge (or even record it on tape), and give them time to do so (see Section 4.3, page 49).

Accidents happen on the plant, not in the design department, and if a report recommends a change in operating rules or a modification to existing equipment, it is usually carried out. If it recommends a change in design codes, it may not be carried out, for several reasons:
- Nobody told the design department.
- Nobody consulted the design department which then resented being told what to do, or found it impracticable. Any recommendations affecting the design department should, of course, be discussed with them before the report is issued.
- The design department had no formal procedure for incorporating the recommendations made in accident reports into their codes and procedures. Someone should be made responsible for reviewing accident reports and seeing that recommendations are discussed with the works and that codes and standards are changed when necessary. Some recommendations may have to be followed up at industry level.

Section 2.1 (see page 4) illustrates the need for these actions. Slip-plating, often recommended after an accident, may have lapsed because the design department continued to design pipework which was too rigid to take a slip-plate.

10.4 FINDING OLD REPORTS

Making old reports more accessible will help us recall the lessons of the past. Suppose someone half-remembers an old report, or wonders if particular chemicals or equipment were involved in accidents. Even in large and sophisticated companies the answer may be hard to find. I received a letter some years ago from a major international company. The writer said, '... I looked for records of accidents in our company for your response and was quite surprised to find that they have been piled up in the stacks for years without any arrangement. No-one in the safety section has been aware of the value of them. We have to depend on the memories of individual persons to find out what happened'.

There is no excuse for this sort of thing in these days of computerised data bases. There are many excellent collections of published information but internal company reports may still be hard to find, and these are often the most valuable.

I suggest that each works and each safety department, perhaps each person, should develop a departmental or personal data base of information that will be useful in the future. It should be a mixture of abstracts of company and published accident reports and recommendations, supplemented by reports obtained from informal contacts with other companies. When the time comes to leave the job, it will be a valuable present to pass on to our successors. Leave your knowledge, not your problems, to your successor.

While keywords make searches quick and easy they have two limitations. The first is that the choice of keywords reflects our current interests but in the future we may have different ones. We should therefore use a data base system with freetext search — that is, the computer can read all the abstracts looking for particular words or phrases.

The second limitation is more serious: when drawing up lists of keywords authors and indexers often miss concepts. For example, a paper carried the title 'Manganese mill dust explosion'[5]. People interested in dust explosions, especially those involving metals, will have read it. They may have found it from one of the abstracting publications or by a computer search; the keywords were 'explosion', 'dust' and 'manganese'. Most other readers will have looked at the title and decided it was not for them. If so, they will have missed several important messages that were included in the paper but have little to do with dusts or explosions.

The explosion occurred because a screen became clogged with manganese dust. Before clearing it, the work crew isolated the power supply by means of an emergency switch. Unknown to them this switch also isolated the power supply to the nitrogen blanketing equipment and the equipment which measured the oxygen content and sounded an alarm if it became too high. Air leaked into other sections of the plant, was not detected and an explosion occurred.

A general message therefore is *when you isolate equipment for repair, make sure you don't isolate other equipment, especially safety equipment which is still needed.* A second general message is *do operators and maintenance workers understand how equipment works?* They may know how each piece of equipment works but it is also important to know how different items of equipment are linked together; that is, how the system works.

Two more general messages, more familiar ones, also come out of the report. Air entered the equipment because *a blind flange had not been inserted* and the screen became clogged because it was finer than usual. Changing the screen size was a modification and its consequences should have been considered before its use was approved.

It is easy to find references to 'manganese' or 'dust explosions' in data bases. If we look for concepts such as 'inadvertent isolation', 'knowledge of equipment', 'isolation for maintenance' or 'modification' we may find little or nothing and may miss important papers. When preparing summaries, indices or lists of keywords or carrying out searches, we often miss concepts because:

- They are imprecise. Manganese has no other name but 'isolation for maintenance' might be called 'preparation for maintenance' or 'disconnection for repair', 'modification' might be called 'change' or 'alteration' and so on.
- The conceptual message or idea is usually less obvious than other messages and the indexer, even the author, may miss it. The author may not even have used the word we are looking for — he may call a modification a change or replacement — so that a freetext search cannot find it.

I have described this point at some length as unless we make a deliberate effort to include conceptual terms in our keywords, or at least in our summaries, we shall find it difficult to retrieve the references we want.

Finally, computerised information storage and retrieval systems will not be widely used if we have to visit the library or seek the help of an information scientist every time we want to use them. They should be accessible from everyone's desk.

10.5 A FINAL NOTE

I have, I hope, established the main thesis of this book — that organisations have no memory, at least so far as safety is concerned — and I have suggested some ways of improving the corporate memory. How far is the thesis true of other functions?

The only other functions of which I have direct personal experience are design and production and it is clear that there is nothing special about safety. Changes made to overcome operating problems are also forgotten after the passage of time and the problems recur, as some of my examples, like those in Section 4.2 on page 45, show.

I have little knowledge of research, marketing, personnel, accountancy and other functions. Perhaps they never forget the lessons of the past. I doubt it. The common factor is people and they are much the same everywhere. Busy with today's problems, they fail to record (in an accessible form) and pass on the information that will prevent tomorrow's problems.

My recommendations apply to production and engineering. Problems of all sorts, not just accidents, will recur less often if we publicise, discuss and record the actions taken in the past, remind people of them from time to time, do not make changes unless we know why the procedure or equipment was introduced and carry out the other suggestions made earlier in this chapter. In other functions different actions may be necessary. I cannot make recommendations but at least I have, I hope, alerted you to the problem.

REFERENCES IN CHAPTER 10
1. Carr, E.H., 1990, *What is History?*, 2nd edition (Penguin Books, UK), 71.
2. Tilley, M., August 1992, *Health and Safety at Work*, 14 (8): 32.
3. Institution of Chemical Engineers, Rugby, UK, various dates, *Hazard Workshop Modules*.
4. Amoco Oil Company, Chicago, Illinois, USA, 1984, *Process Safety Booklets*.
5. Senecal, J.A., October 1991, *Journal of Loss Prevention in the Process Industries*, 4 (5): 332.

AFTERTHOUGHTS

The following has been adapted from Benjamin Franklin, writing in 1758 (changes in brackets):

'Friends and Neighbours, the [impositions of the Health and Safety Commission] are indeed very heavy, and if those laid on by the Government were the only ones we had to pay, we might more easily discharge them; but we have many others, and much more grievous to some of us. We are taxed twice as much by our Idleness, three times as much by our Pride, and four times as much by our Folly; and from these taxes the Commissioners cannot ease or deliver us.'
(The original is quoted in *Economic Briefing,* September 1992, No 4, 7 (HM Treasury))

Many primitive peoples believe that their ancestors will return to haunt them if they do not remember them.

'Behind this idea lies something more than a mere fear of the poltergeist. The notion that the untended dead haunt the living in the form of spooks or spectres is simply a symbolic way of saying that if the past be forgotten or ignored and the connection with it negligently dismissed, it will nevertheless rise up of its own accord and obtrude itself upon the present ... if the past be interred and forgotten, its grave will be unquiet; only when it is fully integrated with the present will it cease to behave like a restless ghost.'
T.H. Gaster, 1980, *The Holy and the Profane* (Morrow, New York), 191

'Safety is not an intellectual exercise to keep us in work. It is a matter of life and death. It is the sum of our contributions to safety management that determines whether the people we work with live or die.'
Brian Appleton, Technical Assessor to the public inquiry on Piper Alpha

In many of the incidents described in this book they died.

INDEX

A

Abbeystead	41–42
Aberfan	43–44, 116, 163
abrasive wheel	59
abstracting	167
accelerating rate calorimetry	127
access	16, 54, 102, 145
accident causes	107, 163
accident investigation	97, 106–109, 116, 120, 168
accident reports	1, 2, 106–109, 120–121, 140–141, 166, 171
acetone	143–144
acetylene	124
acids	58
actions, causes of	70–90
added-on protective equipment (see inherently safer design)	
advisers, relationship to managers	94–95
aerosols	164
agitation	46–47
aircraft	52, 140
alarms	13, 34, 74–76, 94, 98, 101, 104, 135–136, 154–155, 158–159, 169, 172
Alkali Act	63
Alton Towers	135
aluminium	108
ammonia	28
anonymity	67
Armagh	137
asphyxiation	138
assessment of hazards	82–83, 103–105
asymptotes	81, 92
attention lapses	18–22, 109, 122, 139, 158–159
audits (see also inspections, checks, tests, surveys)	50, 96–101, 109, 120, 130, 151, 161, 170
auto-ignition	6, 70
automatic train control	122–123

B

BANANA	87
Bhopal	64, 84, 104, 114–115, 161, 163
black books	170
blame	66, 67-68, 106, 116–117, 144
blanks	9, 28
blast walls	81
BLEVEs	9, 76–80, 114, 154
blinds (see slip-plates)	
blowdown lines	53–54
bolts, failure of	50
breathing apparatus	10, 48, 55
brittle fracture	140
buildings	10–15, 41, 144, 154, 156, 164
burn injuries	145
butane	9, 74, 143
butene	143

C

cameras	148
carbon disulphide	124
carbon monoxide	48
carelessness	139–140
catalysts	113, 143
causes of accidents	107, 163
caustic soda	143–144
change	70–90

INDEX

encouragement of	88
excuses for avoiding	89
gradual	99–100, 112
influence of individuals	87
check lists	99, 102
checks (see also audits, inspections, tests)	109, 120, 135, 149, 152, 166
Chernobyl	162–163
children	137–138
chlorine	67
chokes	15–18, 53, 154, 172
Clapham	56–57, 115
codes	21, 104, 136, 171
collapse of suspended load	48
collapse of walkway	132
colliery tips	43–44
common mode isomorphism	45
communication	21, 95–96, 116
company organisation	85
compasses	111
competency	41–43
complication	115
compressed air	9, 16–17, 21, 32, 154, 160
compressors	11–14, 50, 52, 154, 164
computer control	164
confidentiality	66
confined spaces (see entry)	
congestion	9
construction	130–131
construction materials	58–59, 78, 108, 111, 134
contractors	47, 55, 111, 125, 128–132
corrosion	42, 52–53, 58–59, 145
cranes	48, 136
crapgrinders	24–26
creep	141, 155
criteria for safety	81–82, 92, 104
criticality	10
crowd control	137
Crusaders	131
culture	76, 84, 88, 94, 131–132, 150, 169

'custom and practice'	76, 93–94, 129, 150

D

Daedelus	140
dangerous occurrences	148, 166
Darby	151
data bases	172
Dead Sea Scrolls	2
de-briefing	49, 171
decomposition	45
degrees of freedom	136
demolition	111
design	22, 24, 41–42, 71, 84, 94–95, 108–109, 112, 114–115, 132, 139, 149, 158–159, 171, 173
design limitations	140–141
diesel engines	54, 86, 101
difficult procedures	109, 113
discussions	22, 73, 150–151, 167–170
dispersion	78
dog allowed one bite	104
doubts	42, 43
drainage	79, 80, 101, 103, 128, 138, 144
drain holes	101
drain lines	52–53
drums	48
dust	46, 98, 172–173

E

earthing	101, 108
electrical area classification	79
electric shock	49
electron beam accelerator	135
emergency equipment	101, 136–137
emergency procedures	56
entry	5, 47–48, 55
equations	62
equipment	62
abandoning	22

177

knowledge of	172–173
old	11
removing	22, 144, 170, 174
temporary	41–42, 48
escalators	56
ethylene	9, 11–13, 44–45, 81–83, 124
ethylene oxide	81–82, 127
evacuation	13
excess flow valves	77, 79
experiments	62
experts, limitations of	42
explosions	4, 7, 9–15, 20–21, 24, 34, 41–42, 45–47, 49–50, 52, 54, 59, 111–112, 126, 144, 148–150, 152, 164, 169, 172–173
explosives	100, 124

F

Factory Inspectorate (see also Health and Safety Executive)	77, 80, 96, 98
failure mode and effect analysis	104
fatigue	10, 50
fault tree analysis	103–104
feeding stuffs	126
ferries	17
Feyzin	76–80
filters	19–21
fire-fighting	78, 101
Firemaster	125–127
fires	6, 9, 31, 45–46, 49, 55, 56, 57–58, 68, 70–71, 74, 76–77, 86, 104, 108, 126, 148, 149, 150, 152, 154, 169
flame arrestors	27, 52, 98, 101, 166–167
flameproof electrical equipment	12, 49–50, 99, 129
flammable gas detectors	11–12, 30, 77, 80, 98–99, 150
flammable mixtures	4–5, 13, 41, 111, 113, 124
flare systems	10–11, 79
flare stacks	79, 149

flexibility	109
Flixborough	42–43, 64, 83, 85, 104, 107, 112, 114, 116, 130, 156, 161–163
floods	129
foam-overs	34, 98
foams	45
food canning	112
forgetfulness	136–137
formaldehyde	127
formic acid	47–48
fossils	2
freeing from hazardous materials	111–112
furnaces	10, 141, 155, 169

G

gasholders	4–5, 148
gaskets	14, 49
gasoline (see petrol)	
glass	151
gold	1

H

handover	51
Hastelloy	58–59
hazan (see hazard analysis)	
hazard analysis	82–83, 95, 103–105, 150, 158, 162
hazard and operability studies	42, 95, 99, 102–103, 128, 130, 151, 161
hazards	
assessment of	82–83, 103–105
identification of	102–104
re-assessment of	127
recognition of	17–18, 99–100, 125–127
hazard studies	104
hazop (see hazard and operability studies)	

Health and Safety Executive	82, 87, 104, 127, 162
Herald of Free Enterprise	115, 127, 135–136
Higee process	85
high integrity protective systems	81–83, 88
Hillsborough	137
hired equipment	101
Hiroshima	48
history	2–3
hoses	9, 29, 31–32, 79
HSE (see Health and Safety Executive)	
human error	22, 68, 107–110, 116, 120, 122–123, 139, 148, 158–159, 163, 169
hydrogen	4, 7, 13–14, 49, 124
hydrogenation	15
hydrogen chloride	63
hydrogen sulphide	17

I

Icarus	140
identification of hazards	102–104
identification for maintenance	33, 111–112, 152
ignition	4–5, 6–7, 10, 12–13, 41, 49–50, 54, 56, 70, 78, 80, 86, 107, 111, 132, 149
impartiality, lack of	138
inadvertent isolation	172–173
inexperience	10
information	
availability of	167
retrieval	22, 170–172
storage	120
inherently safer design	83–85, 87, 103, 107, 114, 121, 159–160, 162–163
inhibitors	29
injectors	10–11, 50
innovation	70–90
encouragement of	88

inspections (see also audits, checks, tests)	12, 18, 31, 43–44, 56–57, 73, 101–102, 128, 149, 154, 156
instructions	21, 96, 97, 109, 128, 139, 149, 171
instruments (see also alarms, tests, trips, high integrity protective systems)	105, 143–144
insulation	77–78, 102
insurance	86, 96, 106, 134
intention, change of	7, 71
interlocks	20, 32, 135, 139
International Safety Rating System	99
inventions	42
inventory reduction	77, 83–85, 87, 114, 116, 121, 159, 161
iron smelting	151
isolation for maintenance	4–10, 54, 70–74, 152, 172–173
iso-octane	74, 143–144

J

Jerusalem	138
joints	11–12, 49–50, 54, 124, 129

K

Kansas City	132
keywords	172
King's Cross	56, 95, 115
knowledge of equipment	172–173

L

labelling	56, 101, 110, 127, 139
Lagos	132
lapses of attention	18–22, 109, 122, 139, 158–159
lapses of memory	136–137
law	86
lawyers	64–68, 116
lay-out	115, 156

lead	1	mixing	46–47
leak detection	154	modification	17, 22, 77, 97, 98, 110–113, 116, 125, 132–133, 156, 169–171, 173
leaking valves	4–10		
leaks	10–15, 21, 49–50, 56, 58, 107, 108, 112, 124, 128, 136, 144, 152, 159	morale	131
lectures	167–170	mounding	77
level glasses	98	myths	163
level measurement	110		
lighting	5, 6		
lightning	108	**N**	
limitations of experts	42	natural gas	86
liquefied gases	9–10, 76–80, 83, 98, 108, 145, 154	near-misses	106
		New Zealand	14
Llandudno	134	NIMBY	87
locks	7, 9–10, 71, 88	nitrates	42–43
Loss Exposure and Technical Safety Audit	99	nitration	46–47
		nitrogen	32, 101, 143, 149, 172
lost-time accident rate	115, 121, 145, 150	noise	11, 14, 93, 99, 164
lost-time accidents	143	non-return valves	101
LPG	10–11, 76–80, 83, 98, 99, 108, 145, 151, 154	non-sparking tools	112–113, 124
		novelty	86
		nuclear power	10, 52–53, 82, 87, 160–161
M		*Nutrimaster*	125–127
maintenance (see also preparation for maintenance)	4–10, 12–13, 22, 33, 49–50, 54, 70–74, 83, 97, 108–109, 114, 131–132, 152, 169		
		O	
		Oak Ridge	10
management	97	oil (see also petrol, olefins)	55
senior	115–116	platforms	9, 30
management of safety	93–121, 148	old equipment	11
Manchester airport	45	olefins	49
Manchester Ship Canal	132	oral history	49, 171
manganese	172–173	overloading of supervisors	131–132
materials of construction	58–59, 78, 108, 111, 134	overpressurising	54
		oxidation	81–83, 127–128
memory	22–23, 166, 170, 172	oxygen	81–83, 172
lapses of	136–137		
memory books	170		
methane	41–42		
methods of working	109, 138	**P**	
methyl isocyanate	85	patience	88, 106, 113
Mexico City	115	peripheral operations	100

180

INDEX

permits-to-work	7, 9, 18, 33, 73, 76, 83, 98, 131–132, 151
persistence	88, 150
petrol	10, 86, 111, 130, 148, 162
phenols	144
phosgene	130
Pilkingtons	151
pipebridges	5
pipelines	4, 33, 78, 86, 128
rupture	35, 44–45, 155
Piper Alpha	9, 104, 116, 126
policy	67, 93–94, 120, 155
pollution	63, 77, 104, 164
polyethylene	9, 67
polymer	29
powder	46
power supplies	104–105
preparation for maintenance	113, 116
pressure, injuries caused by	18–22
pressure groups	87
pressure systems	101
pressure/vacuum valves	28
pressure vessels, opening of	18–22
problems, recognition of	74–76, 81–85
procedures abandoned	4–10, 73–74, 90, 144, 170, 174
propane	21, 68, 76–80
propylene	10, 13, 50, 136, 145
prosecution	68, 117
protective clothing	145
protective equipment (see also alarms, instruments, trips)	101–102, 107, 114, 122, 133–134, 148–149, 159, 161, 167
publication	62–68
encouragement	67–68
procedures	66
reasons against	64–65
reasons for	63–64
publicity	65–66
public reaction	63–66
pumps	6–7, 9–10, 28, 41–42, 46, 50, 70, 79, 80, 99, 102, 108, 138–139

R

radioactivity (see also nuclear power)	110
rail tankers	32
railways	18, 32, 45, 48, 56–58, 81, 122–123, 137
railway sleeping cars	57–58
rapid ranking of hazards	128
reactions, runaway	46–47, 51, 112
reactors	52
'reasonably practicable'	103
re-assessment of hazards	127
refrigeration	68, 77–78
reliability	54, 84, 105
relief devices	48, 98
relief valves	9–11, 28, 76, 78, 92, 101, 105, 139, 154
remotely operated valves	11–12, 32, 77, 79, 99, 102, 128, 136, 154
repairs (see maintenance)	
research	1–2, 45–46, 85, 87, 143
responsibilities	94, 115, 120, 126–127, 130
risk analysis	82–83, 103–105, 158
risk assessment	82–83
risk management	93
road accidents	162
road tankers	21, 28, 32, 86, 101, 111
overfilling	31
runaway reactions	46–47, 51, 112
rust	28

S

safety officers	145, 150, 151
safety professionals	94–96, 149–164
salesmanship	88
sampling	15–16, 79, 98
San Francisco	134
scale-up	112
seals	52–53, 79, 99
segregation	115
Seveso	64

181

ships	47–48, 111, 140	threads	32, 133
signal/noise ratio	99	Three Mile Island	107, 160, 163
simplification	109, 115–116, 159	toilets	17
Sizewell	160	tolerable risk	82
slip-plates	5, 7, 9, 22, 54, 71, 88, 90, 131–132, 152, 171	torpedoes	112
		trace heating	16, 102
slips	18–22, 109, 122, 139, 158–159	training	12, 22, 47, 50, 56–57, 73, 96, 97, 109, 116–117, 121, 131, 135, 138, 143, 149, 151, 160, 167–170
'smoulderings'	56		
sodium chlorate	45–46		
spades (see slip-plates)		tramways	134
spark-resistant tools	112–113, 124	transformers	49
spillages	31–33, 86	transport (see rail tankers, railways, road tankers, ships)	
on water	132		
standards	21, 171	trip failure	30–31, 54
static electricity	151	trips	74–76, 81, 90, 98, 101, 114, 133, 151, 154–155, 159, 169
steam	19, 32, 35, 100, 103		
curtains	11–12, 80, 102	random failure	31
engines	105	spurious	134–135
traps	102	trip testing	29–31
stock reduction (see inventory reduction)		turbines	30–31
stories	167		
stress corrosion cracking	42		
successes	58–59	**U**	
Sunderland	137–138	underlying causes	106–107, 120–121, 122, 125, 129
supervision	47, 56–57		
surveys (see also audits)	50, 73, 75, 97–98, 151, 154–155, 161	uranium	10
		user-friendly plants	109, 163
suspended objects	48		
symptoms, treating instead of causes	18		
		V	
		vacuums	27–29
T		valves	4–10, 15, 30, 49, 54, 58, 70–71, 137
tanks	16–18, 27, 34, 47–48, 55, 68, 148, 154		
		excess flow	77, 79
overpressuring	37–40	non-return	101
sucking in	27–29, 37–40, 166, 169	relief	9–10, 11, 28, 76, 78, 92, 101, 105, 139, 154
Taunton	57–58		
temporary equipment	41–42, 48	remotely operated	11–12, 32, 77, 79, 99, 102, 128, 136, 154
terephthalic acid	46		
tests (see also audits, checks, inspections)	48, 57, 74–76, 81, 83, 90, 97–98, 101–102, 111, 114, 151, 154, 167	ventilation	10–15, 55, 102, 111, 154
		vents	15–21, 27–29, 32, 40–41, 47, 52, 78–79, 98, 102, 128, 139, 151, 154, 164, 167

vessel entry (see entry)		water hammer	35, 100
vessel rupture	9, 16, 18, 47, 54, 76–77, 140–141, 154, 169	water sprays	102, 124
		water supply	41–42
vibration	96	welding	7, 111, 131–132
Victoria and Albert Museum	129	Wigan Pier	104
vinyl acetate	81–82, 127	wiring errors	57
vinyl chloride	67		

W

warehouses	45–46
water cooling	77–78

Z

zoalene	127